Spores

Maria Lodovica Gullino

Spores

Tulips with Fever, Rusty Coffee, Rotten Apples, Sad Oranges, Crazy Basil. Plant Diseases that Changed the World as Well as My Life

 Springer

Maria Lodovica Gullino 🄳
Agroinnova
Università degli Studi di Torino
Turin, Italy

ISBN 978-3-030-69997-0   ISBN 978-3-030-69995-6   (eBook)
https://doi.org/10.1007/978-3-030-69995-6

Translation from Spore, Daniela Piazza Editore, by Stefania Antro

Drawings by Carlotta Bianco

This Springer imprint is published by the registered company Springer Nature Switzerland AG
The registered company address is: Gewerbestrasse 11, 6330 Cham, Switzerland

To my father Gino and uncle Piero

Pater ipse colendi
haud facilem esse viam voluit,
primusque per artem
movit agros, curis acuens mortalia corda
nec torpere gravi passus sua regna veterno.
Ante iovem nulli subigebant arva coloni;
ne signare quidem aut partiri limite campum
fas erat; in medium quaerabant, ipsaque tellus
Omnia liberius nullo poscente ferebat.

The Father himself hardly willed
that agriculture would be easy
when he called forth the field with his art,
whetting human minds with worries,
not letting his kingdom slip into full-blown
laziness.
Before Jove took power, no settlers broke the
fields with their plows;
it was impitous then to mark the land and
divide it with boundaries;
people sought land in common, and Earth
herself
gave everything more freely when no one
made demands.

Virgil, Georgics, I, 121–128.

Foreword by Francesco Profumo

University professors are often considered isolated in their ivory tower, avoiding to "get their hands dirty". It is certainly not the case of the author of Spores, who is a passionate researcher, is very close to the territory, and has always sought to combine basic and applied research to respond to stakeholders' requests.

Researchers do not always feel the need to disseminate their results. In this volume, which is the new, international edition of a previous book, published in 2014, Maria Lodovica Gullino goes further, trying to tell the fascinating story of some plant diseases that have deeply affected the economy of countries throughout the world. I, therefore, greatly appreciated the reason that prompted the author to present the discipline she is dealing with in a lighter form, trying to explain to a wider public the social role of plant pathology and the impact that some plant diseases had on our history.

The availability of food for the whole planet and the beauty of the landscape that surrounds us rely on plant health. This is what Maria Lodovica Gullino reminds us with *Spores*, leading us—with an accessible language—on a long journey to discover plant diseases of the past and more recent ones, highlighting the social and economic consequences they had and might still have in the future.

The recent Covid-19 pandemic showed us the importance of environment and plant health. This new version of *Spores* has indeed included the new, interesting vision of circular health.

I also enjoyed the personal cut of this book, reporting author's life experiences, which gives the reader a very lively cross-section of the world of research with its laboratory and field trials and the portrait of colleagues from all over the world.

My hope is that *Spores* may bring young people to discover the fascinating world of scientific research and convince all of them of the importance of investing resources in plant health. Investing in plant health means investing in planet health, thus in our own future.

Francesco Profumo
President of Fondazione Compagnia di San Paolo
and former Italian Minister of Education, University and Research
Turin, Italy

Acknowledgements

For this 2021 English edition, I must thank my Italian publisher and friend Daniela Piazza who encouraged me to write *Spore* in 2014. Thanks to my American friends who suggested me to consider an English edition. Thanks to Stefania Antro for her careful translation. Thanks to my colleagues and friends and to the many people who, completely unknowingly, are part of this book. Thanks also to the many plant pathogens that made my life and work so interesting.

Introduction

Why This Book

Plant diseases caused, in the past, significant economic losses, deaths, famine, wars, and migration. Some of them marked the history of entire countries. One example among many: the sudden development of potato late blight in Ireland in 1845–49.

Today plant diseases are still the cause of deaths, often silent, in developing countries, and relevant economic losses in the industrialized ones.

This book, written with much passion in its Italian version in 2014, does not want to be a Plant pathology text. On the contrary, it wants to describe, in simple words, often enriched by my personal experience, various plant diseases that, in different times and countries, cause severe losses and damages. This English 2021 version is not just the translation of the first edition, but an updated version, with a much more international perspective.

Plant diseases did cause not only significant economic losses but, often, also relevant social effects. Just consider that one and half million people in mid-1800 left Ireland, moving to the United States and Canada as a consequence of the Irish famine caused by late blight of potato, which destroyed a crop that was fundamental to their diet.

Besides the so-called "historical plant diseases", in the process of writing this book, I wanted, by letting take my hand, carried away by my past, to describe also some diseases that though not causing famine or billions of losses, because of their peculiarity, might be of interest for the readers.

One-hundred-seventy years after the development of late blight of potato in Ireland, despite the astonishing results obtained in biology, in general, and in plant pathology, in particular, plant diseases can still cause enormous economic losses in industrialized countries and deaths and famine in third countries. Diseases not only affect agricultural crops, but also ornamental plants and forests. In this case, the damage is most aesthetic than economic and can lead to profound changes in the landscape. Without mentioning the risk represented by the fall of trees, which, by happening in open public spaces, can cause damages to people and things.

Moreover, some pathogens, because of their nature or their way of spreading, can cause anxiety and fear.

This book has the aim of describing, though in a quite light way, the social role of plant diseases, letting the reader know the topical importance of the discipline that studies them, plant pathology, as well as the role of plant pathologists in our society.

Thus, this book has not been conceived and written for experts, but for a broader audience, willing to learn more about plant health and to understand the reasons why so many people, in the past and nowadays, choose to be plant pathologists. This is because plants produce most of the food that we consume, which we expect to be healthy and safe, and also because plants make the world beautiful. By reading this book, the readers will understand why plant pathologists are happy and humble people, very conscious of the social importance of their daily work.

The title "Spores" is evocative of the reproduction means of fungi. Spores are small, light structures, often moving fast. The chapters of this book will be short and concise. Like spores! Plant pathologists reading this book should forgive the fact that, sometimes, also pathogens such as viruses and phytoplasms, not producing spores, are described. Moreover, sometimes I dwell upon some curiosities which, at first reading, seem to have little to do with spores!

At the end of the book, a glossary will help readers when wording becomes too complex.

The first part of the book (Ancient spores) describes quite a few historical famines, together with some episodes that will help understanding how some plant diseases did affect human history. This part of the book will also cover diseases that contributed to modify the landscape in many countries.

The second part of the book (The spores of my life) tells, in a much more personal style, about more recent plant diseases, sometimes very peculiar, which have been the object of my work or that have been important during my working life. Very contemporary plant diseases! This part of the book also includes many characters: researchers, students, people I met in different parts of the world, during my work and my many travels. And also people who in some way have been related to my work. Why so many real people in this book? My intent is to give the readers a cross-section, very much lived, of the world of research. This part of the book also deals with plant protection. Because, at the end, plant pathologists study plant diseases with the final aim of curing them, exactly as a medical doctor. As a plant pathologist, I would describe myself as a plant doctor, doing research my entire life in order to prevent plant diseases and to manage pathogens with methods having a very low environmental impact. As a plant doctor, I always tried to adopt strategies and products "alternative" to the most traditional chemical ones. Whenever possible, of course.

The third part of the book deals with Scary spores: those pathogens that suddenly emerge, in different parts of the world, causing economic losses and destroying, in our century, entire crops. Such plant pathogens do not cause anymore famine, at least in industrialized countries. However, they make us feel helpless and generate fear. This part of the book also talks about pathogens used to destroy coca crops, and about pathogens that might be used against our crops, in a biological warfare. This section of the book also aims to tell readers how plant pathogens are and must still be considered a danger.

The book ends with a look into the future. What is going to happen in the future? A glance inspired by research, but also using some imagination. The recent Covid pandemics did shuffle the cards in any field, including our discipline. What did we learn from it? Probably it is too early to say. However, the first impression, also on the wave aroused by the International Year of Plant Health, celebrated in 2020, the very same year of the Covid pandemics, is that the focus in the future should indeed be more on plant health than on plant diseases.

With the full awareness that sometimes stories are repeated. Thanks to God, maybe... If not, what would plant pathologists be needed for?

Something Very Personal

"You are a plant pathologist? So weird! I never met one before!" Yes! I'm a plant pathologist! Once, every time I was questioned about my job, I did answer with a generic "I'm a researcher". Later on, I started better specifying my research field. But it was only a few years ago that I discovered how fascinating was to my interlocutor to meet with a plant pathologist. At the beginning I was really surprised, since there are so many plant pathologists worldwide. At least 50,000 of them belong to the International Society for Plant Pathology (ISPP) that I had the honour to lead from 2008 to 2013. We are the plant doctors! And we are an army!

Why did I choose this job? Actually, maybe it's this job that chose me.

As a daughter of agricultural entrepreneurs, I was born and grew up in the tiny and beautiful town of Saluzzo, in northern Italy. In the late spring, every year we left our house located downtown, to move to the countryside, and we spent the summertime in the so-called "Ciabòt" (a dialect word meaning a small farm), until the beginning of school. There was only one mile as the crow flies between the two houses, but what a difference! A whole other life. I can really say I grew up in the green, between meadows and orchards. My father, a graduate in Agricultural science, as well as my mother, was a very innovative entrepreneur and my first mentor. He passed on to me his passion for agriculture. An agriculture very much environmentally sound, though oriented to productivity, by taking advantage of the most advanced technologies. In the late 50s-early 60s, I learned from my father how to introduce the most advanced technologies making the better and safer use of them. The agriculture that I learned since then is very much sensitive to consumer's health, grower's safety and environment's protection.

I still remember with emotion, after more than 50 years, my father's recommendations to farm workers on how they had to be careful handling pesticides.

In my second grade school, when I was a seven-year old, in the first compositions that we had to write, I was already mentioning pesticides (DDT, particularly), of which I knew, thanks to my father, the technical name (dichloro-diphenyl-trichloro-ethane) and its mode of action.

The deep knowledge—always sustained by continuous reading of scientific articles—that my father had about plant pathogens and their epidemiology, pesticides, their mechanisms of action and their positive and negative characteristics, did strongly influence my future choices, although I did not probably realize it. Peach trees were the passion and the proud of my father. Harvest time was eagerly awaited. All the photos of my childhood include peach trees and/or peaches. In fact, as my mother used to say, my father photographed peaches with me in the background.

What my father, and me as a consequence, feared most at the time, were the storms (or worse the hail) and, among plant diseases, brown rot of peach and apple scab. Even today a thunder is enough to bring me back to the nightmare of a hail storm, which can destroy the work of a year in a few seconds.

I learnt very early from my father almost everything about the main pathogens of fruit crops. He used to show me the symptoms on fruits and leaves, describing the environmental conditions that favoured the development of such diseases, explaining the strategies he would adopt to manage them. I was very impressed by the attention he paid in choosing the most appropriate and effective technical means, always with attention to their environmental impact. Not so common in the late 50s-early 60s.

On summer evenings dad used to drive me and my younger brother to enjoy ice-cream downtown. Later, going back to the Ciabòt, in the dark, he would drive slowly, with the dazzling lights on, around the perimeter of the orchards. Depending on how many *Cydia molesta*, a dangerous peach moth, were crashing into the windshield, my father would decide whether or not to spray the orchards with an insecticide the day after. Here is, simplifying to the maximum, what we now call the concept of "threshold", the level of infestation beyond which it is necessary to adopt control measures.

I miss so much the long chats with uncle Piero, a great researcher who was studying the breast cancer at the highest levels in Bethesda (Maryland, USA). On his summer trips to Saluzzo, his hometown, this seemingly aloof and very serious uncle devoted much time to the curious brat I was. He would enchant me with his stories that highlighted beauty and secrets of the scientific research, instilling in my mind a keen interest for an unknown world that slowly, year after year, became more and more accessible. Therefore, I heard earlier talking about research and its evaluation, about the impact factor (an index that reflects the importance and influence of scientific publications), etc.

When I was five years old, I did not know how to read or write, since I refused to attend the kindergarten, but I was already subscribed to the English version of the National Geographic Society. "Start looking at the photos!" used to say uncle Piero.

Perhaps unconsciously my father and my uncle were raising a plant pathologist.

Influenced by uncle Piero, dreaming of America and eager to go beyond the photos of National Geographic, I accepted with great enthusiasm the proposal of my father to seriously study English since the second grade, with the help of Miss

Sue Chapman, who stayed in our house for more than one year. To her infinite patience I owe not only the fast and easy learning of English, but also my early love for this language that, as I understood later at the beginning of my professional life, would represent for me a formidable tool of work.

Since I was a child I wanted to be a researcher, following the example of uncle Piero, but once at the University, the paternal imprinting prevailed and the passion for botany, along with some fortuitous encounters, directed me towards plant pathology. That's why I am a plant pathologist.

The long summers I spent at the Ciabòt brought me closer not only to nature but also, and I would say especially, to agriculture and farmers.

Having understood since childhood how serious could be the damage caused by insects and plant pathogens, and how, despite being very hard, the farmers' work could be lived with passion, stimulated me to engage in agricultural research, which I intended not as an end in itself, but as a mean to provide some practical solutions.

I do not hide my passion for agriculture. In my many trips around the world, I have always been drawn to countryside and farmsteads, and I consider farmers—passionate, curious and fulfilled workers—the best part of the productive sector.

Today, after more than 45 years of research, I feel the same enthusiasm of the beginning, consider myself a lucky person, love my work and could never imagine to carry out a different profession.

Contents

About the Author

Maria Lodovica Gullino Born in Saluzzo, of which she is very proud, Maria Lodovica Gullino since the late 1970s with great concreteness and passion has been carrying out research on plant diseases at the University of Torino, where she is a full-time Professor in Plant Pathology and Vice-Rector for the valorization of human and cultural resources.

Daughter of agricultural entrepreneurs and an entrepreneur herself, she lived and worked for long periods abroad. When she is not travelling, she lives in Torino, where she directs the Centre of Competence for the innovation in the agro-environmental field, which she established at the University of Torino in 2002. For work, she has travelled the world far and wide, leaving each time with a suitcase half empty and then full of stones—precious or not—and coloured fabrics at the return.

As a journalist, she loves to read and write. After more than 1000 scientific papers and books, including the series "Plant Pathology in the twenty-first Century" (Springer, 2009–2021), she wanted to try her hand at a lighter writing. She started with "Spore" (Daniela Piazza Editore, 2014), followed by a children's book in 2015, "Caccia all'alieno" (Alien hunt), on alien pathogens threatening fruit and vegetable crops, and in 2016 "Valigie: cervelli in viaggio" (Suitcases: brains on the move), edited by the same publisher. In 2018, she wrote "Angelo, il Dottore dei Fiori" (Dr Flower), the story of Professor Angelo Garibaldi, researcher and co-founder of Agroinnova, illustrated by Gabriele Peddes and published by Edagricole, Business

Media. Adapted also for the theatre, this book has offered the opportunity to make even the youngest ones aware of the fascination of research on plant diseases. For the International Year of Plant Health, in 2020, she wrote another book for children, entitled "Healthy plants, healthy planet" (FAO), translated into several languages.

Ancient Spores

This first section of the book, going through a period of several millennia, starting from the first plant diseases mentioned in the Bible, describes some serious diseases of the past that have truly marked the fate of certain countries. The history of those diseases reminds us of the fragility of our production system, and of the crucial role that agriculture plays in every age and society. Moreover, some episodes in particular offer clear proof—if ever there was need—of how history repeats itself. The coffin ships that in nineteenth century carried to North America the Irish immigrants escaping the famine caused by potato late blight in their country resemble the makeshifts vessels leaving nowadays the African continent along the Mediterranean route. Now as then, famine and starvation force entire populations to migrate.

The Tulip Mania—which I will discuss more later—that happened in Holland in the distant sixteenth century, doesn't remind us of more recent financial bubbles?

As history teaches us, plant pathogens have no boundaries and no mercy. They affect indifferently rich and poor countries, important and small plants. Thanks to their spores, they travel from one country to another, at a speed that has changed according to the development of transport. And when pathogens affect ornamental plants that beautify our environment, the damage is mostly aesthetic, rather than economic, but still very important.

The past also teaches us that pathogens can become real weapons, used to starve the enemy, in the so-called biological warfare. As we will see, several countries have owned for years laboratories equipped to produce pathogenic spores of important crops: war spores.

Here then is a series of historical spores, described in short—and hopefully easy-reading—chapters.

Plant Diseases in the Holy and Classical Texts and in the Art

Plant diseases are as old as the plants themselves. Fossils present signs of various pests and severe famines, due to drought, locust invasions and fungal diseases, as reported by several ancient writings.

The first information about plant diseases date back to Assyrian-Babylonian, Egyptian and Jewish writings and figures.

Famines attributed to rust attacks on cereals are mentioned in the first book of the Bible (*Genesis*, 42, 1–5). This disease was well known in the past. Its typical symptoms on cereals—rust-colored pustules—together with the associated remarkable yield losses, made it so popular that Romans thought it was the work of divinity (see box page 5). Moreover, in some passages, diseases likely ascribable to coals, capable of destroying seeds, are reported around 750 B.C.

Among the Greek authors, it was Hesiod the first to deal with agriculture. In his poem Ἔργα καί Ἡμέραι (*Works and Days*) (7th century B.C.), he gives practical advice for the cultivation of fields and suggests the days of the month more suitable for performing certain activities. However, it is with the works of Aristotle and Theophrastus that we have much more precise information on plant diseases. The latter, in particular, author of the two most important texts of Botany of the past *Historiae plantarum* and *De causis plantarum* (*Enquiry into Plants* and *On the Causes of Plants*), speaks of propagation of plants by means of seeds and grafting and of the effect of environmental conditions on plants, and refers to the cereal rusts, also providing basic information on their biology and epidemiology. The Greek writer (371–287 B.C.), in fact—who for the two volumes mentioned above is rightly deemed the "father of botany"—reports in his writings a correlation between the severity of rust attacks and the location of the most affected fields. Although he did not know the causes of this disease, Theophrastus observed, for example, that cereals were more frequently affected by rust than vegetables. Already in 470 B.C., moreover, the Greek philosopher Democritus provided some interesting indications to protect the wheat from rust, advising the use of fumigations.

Later on, various Latin authors, including Cato, Varro, Virgil, Columella and Pliny the Elder, often cited in their writings some fearsome plant diseases able to

© The Author(s), under exclusive license to Springer Nature Switzerland AG 2021
M. L. Gullino, *Spores*,
https://doi.org/10.1007/978-3-030-69995-6_1

cause severe famines. Speaking of Latin writers, it is interesting to note the different conception of agriculture of two great poets, Virgil and Lucretius. For the singer of the Augustan era, the adversities of nature, which men have to overcome, are a gimmick of Jupiter to prompt human ingenuity. In Lucretius, however, it is the sign that nature was not created by divine agency for the good of mankind, since "it is marked by such serious flaws" (*"tanta stat praedita culpa"*, *De rerum natura*, v. 199), although the philosopher himself believes that through work, man has come out of the wild state. In the *Georgics* there is an interesting description of rust, which devours the stems of wheat (see box page 6). And we can find cereal rust cited again by Pliny the Elder.

Despite the observations of these authors, often based on scientific ground, for a long time plant diseases were considered as magic or sidereal phenomena attributed to supernatural forces. It was general belief among farmers that plant diseases were in some way a consequence of divine punishments.

Severe famines due to fungal diseases also occurred in the following centuries, generally trigged by violent epidemics of wheat rust or other cereal rusts that caused serious yield losses followed by reduced stocks, raids and disturbances. A more recent example is the plague of 1629–1631, described by Alessandro Manzoni in the novel *The Betrothed*.

The invention of compound microscope in 1590 and its improvement by Anton van Leeuwenhoeck in 1683 allowed the first observations on parasitic microorganisms. It was initially botanists who made interesting contributions to plant pathology: after all, what is plant pathology if not the study of sick plants?

Some Italian scientists, such as Pier Antonio Micheli, Giovanni Targioni Tozzetti (see box page 7) and Felice Fontana, greatly contributed to plant pathology foundation as an independent discipline. The first experimental proofs of the parasite relationship between plant and pathogen were carried out by Prevost in 1807 on common bunt of wheat (see page 266), while the understanding of the biological cycle of rusts, by Antoine De Bary in 1865, marked the birth of modern plant pathology.

Born within botany, plant pathology was acknowledged in the last century as a scientific discipline in its own. Advances made in plant sciences, microbiology, biochemistry and biotechnologies have strongly influenced this discipline. At the same time, plant pathology has contributed to the development of many other disciplines.

Also numerous works of art have given, in their own way, a contribution to plant pathology. Plant pathologists always stand out from "normal" people because of the special regard they have for pathogens and plant diseases: it is mainly the infected plants or those with symptoms or strange signs to catch their attention. Even when faced with a picture, for example in front of a still life, a plant pathologist will look for any signs of illness—starting with the magnificent tulips, as much mottled as infected by viruses, portrayed by Flemish and Dutch painters in 1500–1600. Rembrandt and Vermeer lived through the years of the Tulip Mania (sort of collective madness unleashed by these fashionable flowers that led to a real speculative bubble—see page 39), and many of the major Dutch artists of the time chose the infected tulips as subject for their paintings. Famous, among others, is the artwork by the German painter Jacob Marrell (1614–1681), active in Utrecht, which portrayed tulips affected by Tulip Breaking Virus.

The Great Hunger (1962) by Cecil Woodman-Smith describes the Great Irish Famine brought in 1840s by the late blight of potato, and so do several paintings of Flemish artists. Among the most famous artwork on this subject, stands out *The Angelus* (1857), initially titled *Prayer for the Potato Crop*, by Jean-Francois Millet, actually displayed in Paris at Musée d'Orsay.

Another very famous work, *The Beggars* (1962), by Pieter Bruegel the Elder, that we can admire at Musée du Louvre in Paris, would portray people whose legs would have been impaired by ergotism (see page 9), a very common disease in those days in Europe.

Going to depth

Prayers to the goddess Rubigo to protect cereals from rust

Plant diseases, the cause of famine, have always been feared and for a long time they were considered a divine punishment. In a certain sense it was believed that God or some deity, by influencing the climatic conditions, favoured the appearance of devastating epidemics.

In the treatise *Naturalis Historia*, Pliny the Elder (23–79 A.D.) and Ovid (43 B.C.–18 A.D.), in the fourth book of *Fasti*, mention the liturgical celebrations (the famous *Rubigalia*) held in Rome during the month of April to appease the wrath of the goddess *Rubigo*, personification of wheat rust.

The ceremony, very well described by Ovid, consisted of a procession of people dressed in white who brought red animals (dogs, cows, foxes) to sacrify as a gift to the goddess, in a forest outside Rome. The origin of these religious ceremonies, practiced for more than 1,700 years, is traced back to the punishment inflicted, with the appearance of cereal rusts, after it had been set on fire with cruelty to a fox. The symbols associated with these ceremonies often presented the reddish and black colours, typical of rust spores, and represented cows, foxes, red dogs etc.

Moreover, in some countries propitiatory processions are still celebrated to obtain good harvests.

Going to depth

The wheat rust at the time of Virgil

Below is how Virgil described wheat rust.

*Prima Ceres ferro mortalis vertere terram
instituit, cum iam glandes atque arbuta sacrae
deficerent silvae et victum Dodona negaret.
Mox et frumentis labor additus, ut mala culmos
esset robigo segnisque horreret in arvis
carduus; intereunt segetes, subit aspera silva,
lappaeque tribolique, interque nitentia culta
infelix lolium et steriles dominantur avenae.*

Virgil, Georgics, II, 147–159

Ceres first taught men to plough the earth with iron,
when the oaks and strawberry trees of the sacred grove
failed, and Dodona denied them food.
Soon the crops began to suffer and the stalks
were badly blighted, and useless thistles flourish in the
fields: the harvest is lost and a savage growth springs up,
goose-grass and star-thistles, and, amongst the bright corn,
wretched darnel and barren oats proliferate.

Magic practices and family matters

The father of modern phytopathology, the Florentine Giovanni Targioni Tozzetti, in his work *Alimurgia* also describes—not without irony and sarcasm—the magic practices that were used by farmers to fight the diseases of plants, such as burning live scorpions or the left horn of a cow. Giovanni Targioni Tozzetti had a son, Ottaviano, who was a physician and a botanist, professor at the University of Pisa. His nephew, Antonio, was a physician, famous for his findings in the field of chemistry, and Director of the "Garden of Simple", now the Botanical Garden of the University of Florence. Antonio married the noblewoman Fanny Ronchivecchi who animated a literary circle in Florence, in via Ghibellina, and who inspired Giacomo Leopardi the *Aspasia Cycle*.

Poisons of the Past: A Very Fascinating Hypothesis

The importance and sad fame of *Claviceps purpurea*, a fungus that attacks rye, are linked to the disastrous consequences for human and animal health caused by the presence of its sclerotia in the flours and fodder. Since ancient times this fungus has been the cause of very serious epidemics of a disease known in the Middle Ages as sacred fire or fire of St. Anthony, now as ergotism (term deriving from French *ergot* = spur, with which in France is indicated the sclerotium produced by the pathogen), responsible for hundreds of thousands of human and animal deaths.

Those who consume food prepared with rye flour contaminated with sclerotia of *Claviceps purpurea* may encounter two main forms of ergotism, probably due to a different ergotoxin content in the contaminated sclerotia as well as to a different amount of ingested ergotoxins: a convulsive form and a gangrenous one.

The first causes very complex pathological manifestations, of nervous type: hallucinations, convulsions, sensation of epidermal burns, abortions, reduced fertility. The second, on the contrary, has a generally acute trend, causing swellings and strong skin rashes of the limbs, which at first assume a purplish-red colour, tending to black, and then going into gangrene due to arrest of the blood circulation, until they become stumps destined to detach from the body of the sick person.

The complexity and variability of the phytopathological picture, its unpredictability together with the total ignorance of its origin, justify the sense of mystery and the superstitions that in the past accompanied ergotism.

The first news of widespread manifestations of ergotism come from France and date back to the end of the 6th century A.D. The disease then reappeared in 857 in France, along the Rhine Valley, causing thousands of victims. Since then, various other intoxications followed: outbreaks of ergotism are reported in 1042, 1066, 1089, 1094.

The foundation of the religious order of St Anthony by Pope Urban II in 1095, for the care of the sick, was one of the few measures at that time feasible to alleviate the consequences of the disease. The monks of St. Antony order used to raise pigs, with whose fat they soothed and cured the wounds of those who, afflicted by skin rashes, turned to the convent to be cured. This is the reason why Anthony the Abbot is often depicted with a pig.

Another interesting fact related to this disease was the finding that many patients with ergotism who went on pilgrimage to churches or monasteries dedicated to St. Anthony of Padua significantly improved or even healed. This led to large flows of pilgrims towards the convents dedicated to the saint who assumed the reputation of holy-healer. Today it is believed—and I don't want to appear blasphemous in reminding this—that it was not miraculous healings but, more simply, improvements due to a change of diet, as sick pilgrims moved from northern Europe, where they consumed rye flour, to southern Europe, where instead they fed on wheat flour, not contaminated by *Claviceps* sclerotia. Preventive measures were introduced only towards the end of the 18th century, after Abbot Tessier had experimentally demonstrated, in 1777, that ergotism was caused by the ingestion of contaminated flours by what we know to be the sclerotia produced by *Claviceps purpurea*. With the spread of this knowledge, fewer and less serious outbreaks have occurred, though they did not cease altogether.

Among the most serious epidemics are those that occurred in Italy in 1789, in Russia in 1926, in Ireland in 1929. A famous ergotism epidemic in 1722 succeeded in blocking the army of Peter the Great who, in order to conquer the Turkish ports on the Black Sea, moved south with his troops. Arrived on the banks of the Volga, he lost his men and animals because of ergotism caused by the ingestion of contaminated flour. The last ergotism epidemic reported in Europe, quickly circumscribed, occurred in France in 1951.

The severity and spread of ergotism were largely due to the strict dietary reliance of many populations on rye. It is believed that ergotism played a significant historical and social role in such situations and was, *inter alia*, one of the main causes, in the years 1670–1745 and 1779–1810, of demographic decline in France, a country where, more than elsewhere, rye remained a staple food, at least until the potato was introduced.

Mary Kilbourne Matossian, historian of Maryland University, studied the role of ergotism in particular and of mycotoxins in general in the history of mankind. In an essay published in 1989—which is strongly recommended reading—she hypothesises that the so-called black death of 1347–1351, the deadliest pandemic ever recorded, caused by bubonic plague, was at least partly due to the immunosuppressant effect of mycotoxins present in contaminated grains. In addition to the serious damage caused in people, in fact, these mycotoxins determined also the death of rats feeding on the contaminated grains. As a result, an increased number of fleas in search of alternative blood sources to that of mice turned to humans and animals, causing their deaths.

An accurate analysis of fertility and mortality in Russia between 1865 and 1914 (when the average diet was very similar to that adopted in Europe before 1750) leads Matossian to correlate the decline of fertility, high mortality and some mental illnesses with the consumption of moldy flour, contaminated by mycotoxins (see box page 12). The historian also convincingly demonstrates how the witch-hunt that affected Europe in the 16th and 17th centuries concerned mainly areas—such as French Alps, central Europe and Rhine Valley—where rye was the predominant cereal crop, and occurred in years characterized by humid climate, favourable to the contamination of rye by *Claviceps* (see box page 13).

According to Matossian, the population growth observed in Europe in the period 1750–1850 had to be linked to a change in diet and the reduction in consumption of rye bread, probably very contaminated, replaced with wheat, maize and potatoes. A sort of litmus test! Matossian noticed that, compared to English, French population alternated phases of demographic recovery and other phases in which deaths due to epidemics of ergotism marked declines in population. When, starting from 1810, French population began to increase, the phenomenon was explained not by an increase in fertility but, rather, by a reduction in deaths.

The toxic fraction of *Claviceps* sclerotia includes, moreover, at least 5 alkaloids whose relative variations of contents can explain the prevalence of one or the other of the different ergotism syndromes described (see box page 14).

Nowadays, even if the pathogen can still be a problem, the modern cultivation and threshing techniques, together with hygienic-sanitary controls, allow to obtain non-contaminated flours.

Going to depth

The methodology used in the study

At present, mycotoxicosis are studied with well-established methodologies, but it has been more difficult for the historian Matossian to study those of the past, lacking any reliable evidence and methods. A nationwide system of cataloguing epidemics began in France in 1776, in 1836 in England and only in the second half of the 1800s in the other countries. Nevertheless, the data from that period must be taken with caution. It was so difficult for the author to study the relationship between the alkaloids produced by the fungus in sclerotia and some bizarre behaviours!

Matossian found a useful source of reference in the diaries of some doctors and some priests, since in the past, as it is important to remind in this regard, many ecclesiastics had a background of medical studies, and therefore were able to provide health care in their communities.

In her approach to mycotoxicosis study, Matossian took into account the incidence of various events, regional correlations between diet and climate associated to fertility and mortality, the exposure to mycotoxins of different social classes and the effect of the elimination of certain products from the diet, such as, for example, the replacement of rye with wheat, maize and potatoes..

The "witch hunting"

It is interesting to note that there have been few cases of witch hunting in Ireland, Spain, Italy and Scandinavia—geographical areas characterized by hot-dry or cold climate. The victims of persecutions were people accused of causing symptoms corresponding to the ergotism classic ones. From a temporal point of view, the peaks of the phenomena have been observed in different periods in the various territories, but always in conjunction with seasons characterized by climatic conditions favorable to *Claviceps* attacks on rye. In the County of Essex, in England, the phenomenon occurred frequently, especially in 1580s–1590s; while witches were generally middle-aged or elderly, the victims were young adults.

Going to depth

Claviceps: a fungus like a pharmacy

Claviceps sclerotia turned out to be a powerful pharmacological mine. Since the 18th century, the dried powdered fungi were used at the time of childbirth to stimulate smooth muscles and thus induce uterine contractions and prevent hemorrhage by exploiting vasoconstriction.

Later on, single *Claviceps* alkaloids or their derivates, found more specific medical and scientific applications that led, among other things, to the discovery of the function and utility of histamine. The hallucinogen LSD (lysergic acid diethylamide) has turned out to be the most effective pharmacological factor of the ergotoxins, present especially in the *Claviceps paspali* species, common parasite on *Paspalum distichum*.

For their pharmacological properties, *Claviceps* sclerotia have been cultured. Production yields have been increased by using strains of the fungus with high alkaloid contents and rye varieties particularly susceptible to *Claviceps* attacks; furthermore, particularly effective techniques of artificial inoculation from mycelium or spores produced *in vitro* have been applied.

Although chemically synthetized, ergot alkaloids are now produced at lower cost from cultivation of suitable *Claviceps* strains on artificial substrates.

Late Blight of Potato: The Great Irish Famine

In 1845, when the potato late blight reached Ireland, potato was a big staple food in the country. Two varieties of this species, selected over time for their high productivity and popularity among consumers, were cultivated on countless acres. In fact, in 1800s, the Irish had developed a complete dependence on this crop, which came from South America, mostly from the region of Lake Titicaca, between Peru and Bolivia. It had been discovered in the 16th century by the Spanish conquerors, who however did not immediately appreciate its enormous nutritional value, as they were much more interested in the gold mining.

The potato was, and still is, a very important crop in South America, especially at the highest altitudes, where maize cannot be grown.

Many finds of the Inca civilization refer to the cultivation of potato, and archaeological evidences attest the presence of this crop as far back as the 400 B.C. In this regard, those who have the opportunity to go to Peru should not miss a visit to the International Potato Centre, a unique experience, which has nothing to envy to the certainly most popular visit to Machu Picchu.

The gold diggers, on their way back to Europe, probably brought some potatoes that suffered the long journey under not optimal conservation conditions. The arrival of the first potatoes in Spain dates back to the second half of 1500s, precisely between 1560 and 1564 at Seville, around 1575 in Portugal and at the end of the century in Madrid. The potato arrived in the British Isles from America in 1586, and after two years was already grown in Ireland.

There was, initially, some confusion between sweet potato (*Ipomea batatas*, very different from the botanical and nutritional point of view) and potato. The name of this species, which was new in Europe, was "translated" in various languages—included French, Dutch and Finnish—as "earth apple".

Potato did not initially have an easy life in Europe. On the one hand, this species was considered more suitable for animal feeding and, on the other hand, being not mentioned in the Bible, it was viewed with some suspicion by the religious world. For many years potatoes were used to feed pigs. There was in fact a certain distrust of what grows underground and it was suggested that its consumption could spread

© The Author(s), under exclusive license to Springer Nature Switzerland AG 2021
M. L. Gullino, *Spores*,
https://doi.org/10.1007/978-3-030-69995-6_3

leprosy. However, the nutritional properties of potato slowly prevailed and hungry people began to appreciate its enormous potential. Besides, in stormy periods, European populations greatly appreciated, among the properties of the species, its ability to grow and produce underground, sheltered, therefore, from raids and war devastations. In 1800s it was certainly a very widespread crop, well suited to Irish climate, and a safe nutritional source.

Although yields varied from one year to another, on average production was sufficient to feed a population that had grown in 1800–1845 from 4.5 to more than 8 million inhabitants. In "good" years, every Irishman ate kilos of potatoes a day. Just potatoes and little else, if not a bit of milk. It was certainly a little varied diet, though capable of providing proteins, carbohydrates, vitamins and minerals, being the potato particularly rich in vitamin C.

While cereals, due to climatic conditions, were cultivated with difficulty, with unsatisfactory yields, enough potatoes to feed a family could be produced on areas smaller than those required by cereals. This long premise was essential to understand the disaster led by the emergence of a new disease, the late blight of potato (caused by *Phytophthora infestans*) which literally destroyed the Irish potato crops in 1845.

In that distant summer of 1845, hot and dry days followed 6 cold and rainy weeks. Potato plants began to lose vigour, presenting showy yellowing followed by darkening and deep alterations of stems and leaves. Under the ground, most tubers were rotten and even those apparently normal rotted very quickly after harvest.

This phenomenon appeared throughout Europe and was particularly severe in Ireland. Why? Firstly because, as already said, Irish diet had become completely dependent on potato. Secondly, what was then a mysterious disease suddenly wiped out all the harvest, forcing people to starvation.

At that time, scientific knowledge was not very advanced. Although the invention of the microscope by Anton van Leeuwenhoek occurred 200 years earlier in the Netherlands, and despite the first descriptions of plant cells by the English botanist Robert Hooke, at the time of the appearance of the potato late blight, spontaneous generation was still held responsible for the presence of "microbes" in altered plant tissues. The cause of this disease was completely unknown. The Irish peasants ascribed its origin to the devil himself, whom they tried to exorcise by spreading holy water on the fields; or to railway engines that, crossing the countryside at the diabolical speed of 20 miles/hour, flooded potato crops with harmful electric discharges (something similar to the present 5G fear!); or to a divine punishment for their sins.

In 1845, when a white mycelium was observed on the affected potato tissues, it was considered the result rather than the cause of this mysterious disease. Certainly a correlation was immediately suspected between the abundant rain which had hit the country on that summer and the emergence of the new disease. But it was not enough. There had been, in fact, rainy years even before, without this leading to such catastrophic consequences. The cause, therefore, had to be sought elsewhere. While, as we shall see, Irish people were starving, many researchers, initially mostly botanists, began to study this interesting phenomenon.

Today the causal agent of potato late blight, once considered among fungal pathogens and now downgraded to fungus-like organism, to put it correctly, is one of the most investigated. But let's go back to Ireland in 1845.

As has been said, the potato disease exploded suddenly, destroying crops all over the country and starving the peasants, who represented the majority of the population. At the same time, late blight attacks reduced the availability of food for the entire country. Very little tubers had left for the sowing of the following year.

The political situation was also delicate: the British refused to help the Irish. And complex market mechanisms came into play, worsening the normal allocation of food (see box page 19). It was not until February 1846 that the United States of America began to send maize to Ireland.

In 1846 it was hoped for a good harvest for a while, but unfortunately climatic conditions were again favourable to the pathogen, aggravating the situation. Starving populations could no longer pay rents (often paid in nature by means of potatoes).

Several paintings by Flemish painters described the famine.

There are hundreds of testimonies of the tragic impact caused by the harvest failures in Ireland: hunger, unrest, raids, illnesses, deaths. Also a famous Irish folk ballad, *The Fields of Athenry*, composed in 1979 by Pete St John, recalls the Great Famine (see box page 21).

Unfavourable weather conditions continued throughout the 1800s (the so-called "years of hunger") with disastrous consequences for those populations whose livelihoods relied on potatoes. Within 15 years Ireland lost 2.5 million inhabitants: one million died of hunger and hardship, one and a half million emigrated, especially to U.S.A. and Canada.

Today it is estimated that 10% of the population of the United States of America originates from starving Irish people who fled their country in those years. In U.S. A. Irish immigrants formed a large colony that would become very influential on political, economic and social matters (the late John F. Kennedy, president of United States from 1960 to 1963, was himself an Irish descendent).

To give an idea of the sufferings experienced by the Irish, it is enough to mention that only in 1847, 100,000 people left the country to reach Quebec, Canada, through inspections and quarantine. Of them, 5,000 died during the trip and 5,000 of typhoid and dysentery, before reaching the much-desired destination (see box page 21). The Boston Irish Famine Memorial by Robert Shure (1998) reminds younger people of that enormous tragedy (see box page 22).

The numbers above testify the serious hardship caused by the famine unleashed in Ireland by the arrival of the potato late blight, an event that has marked the history of a country, raising questions that deserve an answer. That is what we will try to do in the following pages.

Could such a tragedy be repeated today? Perhaps not in the industrialized, but in the developing countries such a tragedy is still possible.

Could such a tragedy have been avoided? Certainly not in 1845. Today, probably yes.

The epidemic induced researchers of the time (especially botanists, as already mentioned) to study plant diseases. It was Rev. Miles Joseph Berkeley, in England, to recognize that the fungal pathogen described and named by Camille Montagne, military physician in Napoleon's army, could be the causal agent of potato blight.

The study of this disease led to a great competition among the first scientists. As already mentioned, at that time the theory of spontaneous generation was prevalent. Even doctors and clergymen, who represented the most evolved and cultured part of the population, seeing the signs of the pathogen in the tubers, on potato stems and leaves, believed they were produced by the plant itself as a result of rot, rather than being the cause of rot and death of the plant.

Berkeley's claims were challenged by the English botanist John Lindley, and The Gardener's Chronicle published their disputes that remain, perhaps, the most profound philosophical dissertations among researchers in the field of plant pathology. Nowadays, few phytopathologists would be able to face a discussion with so much philosophical depth. Yet, unfortunately, those disputes did not manage to keep late blight of potato under control.

It was only the German botanist Anton de Bary in the years after 1860 to demonstrate that this disease was due to an organism, now considered similar to a fungus. He assumed that its spores (scientifically sporangia), observed in great quantity on the infected tissues, were able to penetrate and infect healthy tissues. The same researcher proved that the presence of high relative humidity conditions was not sufficient to unleash the emergence of the disease. In the absence of the pathogen, in fact, it did not cause symptoms. I would like to highlight the importance of the results obtained by A. de Bary, who managed to prove experimentally that the major culprit of the devastating Irish famine was actually an organism, at that time considered a fungus (see box page 22).

So, what happened in the end? Was A. de Bary's research enough to manage the disease? Absolutely not.

It was only in the late 1800s that the discovery of the effectiveness of copper salts against downy mildew of grapevine (see box page 26) made it possible to control *Phytophthora infestans*. The disease became more topical during World War I. Potatoes had, in the meantime, become a big staple food in the German diet. During the war, there was a competition between industry and agriculture for the use of copper, which was employed by war industry to produce weapons and bullets.

In the winter of 1915–1916 many potatoes rotted during storage because of late blight attacks. Hoping for a warmer climate, less favourable to the pathogen, in the following spring and summer, the German authorities chose to use copper for war purposes, but again a climate favourable to the pathogen in spring-summer 1916 caused severe attacks with consequent famine and death of 700,000 Germans among the civilian population. Those who could find them, ate turnips, that's why the winter of 1916–1917 went down in history as the "Turnip Winter". It seems incredible, but a microorganism managed to upset the German army more than the enemies.

Even today, despite the availability of resistant varieties and effective tools employed for the potato late blight management, this disease remains a serious problem and its study keeps alive the interest of a large number of researchers.

Going to depth

At the roots of famine

In fact, as the Indian economist, 1998 Nobel Prize winner Amartya Sen explains well, complex market mechanisms also came into play and, essentially, nobody was concerned in defending the rights of the most vulnerable sections of the population. In addition to potatoes, Ireland produced wheat and meat: in all the years of famine, every day dozens of food-laden ships departed from Irish ports bound for England. For a ruthless market behaviour, wheat and meat yielded higher prices in England than in Ireland. It was a great failure of the Institutions which were supposed to ban the export of commodities.

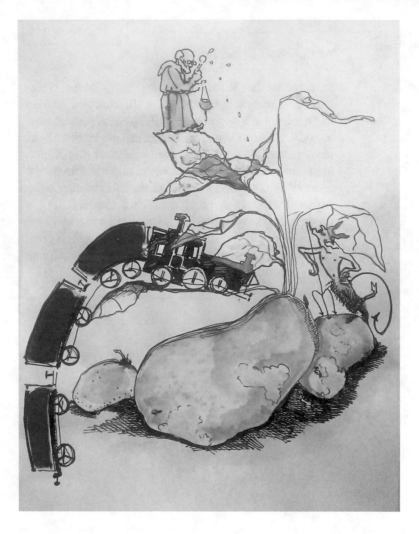

The heart-rending ballad of the fields of Athenry

The lyrics of this Irish folk song feature the story of an Irish convict who through the wall of the prison hears a conversation between a young man from Athenry, Count Galway, where the famine is very severe, and his wife. The young man has been convicted of stealing food for his starving family. He must now be transferred to the Australian penal colony of Botany Bay. The last verse of the ballad tells the departure of the ship with the young convict on board, while the wife remains on the ground. This ballad has been adopted by the Irish national football team supporters since the 1990 World Cup.

Coffin ships

The situation in Ireland, as a result of late blight attacks on potato crops, became so tragic that about one million people emigrated, as has been said, mainly to the United States and Canada, but also to England and Wales. They embarked on crowded dilapidated ships, with poor access to food, drinking water and sanitation—the so-called coffin ships—and there many of them, already weakened by starvation and misery, often affected by typhus or other infectious famine-related diseases, met their end. Today history is repeated, unfortunately, with the makeshifts vessels carrying refugees and migrants from non-EU countries, mostly from North-Africa, towards our continent.

To not forget

The memory of the catastrophe that struck Ireland is still so vivid in the country that in 1997, a sculpture of a small sailing ship with the bow facing the Atlantic, with a series of bronze skeletons under the mast, was placed in the shadow of the Croagh Patrick, sacred mountain for the Celtic people. On the other side of the Ocean, in Boston, where thousands of Irish people fled to escape from the famine, surviving to the epidemic and the long sea journey, the Robert Shure artwork testify to the descendants of immigrants the memory of what many do not hesitate to call a holocaust. I'm sure that many plant pathologists visited it in 2018, while attending the International Congress of Plant Pathology.

Numbers don't lie

Plant pathologists, or better those among them who carry on research on epidemics, refer to a model of disease causation called the "triangle", which consists of three essential elements: a susceptible host, an external agent (a pathogen able to attack the plant), and environmental conditions conducive to infection.

A very simple concept, which is at the root of all diseases. In the case of potato late blight, it was the arrival of a new pathogen to compromise the situation.

American Diseases of Grapevine: Why Americans Drink Whisky

The common grapevine is a very ancient species. Fossils testify to the presence of wild grapevines in several parts of Europe already 500,000 years before Christ, and in 3,500 B.C. they used to eat its fruits.

The European or domestic grapevine (*Vitis vinifera sativa*) is the one grown today to produce wine and table grape. It seems that the first to cultivate grapevine were the Armenians and the Georgians, several millennia B.C., than the crop spread rapidly in the neighbouring regions (see box page 25). In the third millennium before Christ grapes began to be used for winemaking. Wine is also mentioned in the Bible. In Italy the cultivation of European grapevine began with the arrival of the ancient Greeks in Sicily and, more or less in the same period, the Etruscans had begun to grow grapevine in Tuscany and Lazio. It was, however, the ancient Romans who spread grapevine and wine almost all over Europe, in the conquered territories. Grapevine growing continued in the following centuries, representing a rather easy and unpretentious production.

Suddenly, around the middle 1800s, things got complicated, as three new plant pests arrived from America, one in a row to the other, causing very serious damage to European vineyards and making viticulture much more complex. They were caused by two plant pathogens—causal agents respectively of powdery mildew (*Oidium*) and downy mildew—and by an insect pest, the grape philloxera, that were likely present in North America since a long time, even well tolerated by wild grapevine. Because of the presence of such parasites, however, the Americans had never been able to successfully grow *Vitis vinifera* cuttings, which they imported from Europe.

The introduction of the three "American" pests in the Old Continent was probably due to the initiative by some ignorant people who, in the middle of 1800s, tried to cultivate American wild grapevine in Europe.

The first to reach Europe was the agent of powdery mildew, observed in 1845 first in England, then in France, in Italy and in other European countries. Its symptoms were quite serious, until it was found out that sulphur was highly effective against this fungal disease.

Shortly afterwords, it was the turn of philloxera, caused by an insect little harmful on American vines, but very injurious on the European ones: its name, *Phylloxera vastatrix*, well describes its devastating action, especially against the root system. Fortunately for European winegrowers, the grafting of a European grapevine scion onto the rootstock of native American (resistant to the pest) vine cuttings solved the problem. This technique is still commonly used today: new vineyards plantation is always done using grafted plants.

The third pathogen, *Plasmopara viticola*, causal agent of the disease known as downy mildew, was the most insidious. This pathogen had already been described in North America in 1834, where it was probably endemic on wild grapevine, without causing serious damage. The best trained European growers were aware of the presence of this pathogen, known to obstacle the cultivation of European vine on the Atlantic coast of North America. Despite the preventive measures taken, the agent of downy mildew reached Europe, causing the first damages in 1878 in some vineyards in the region of Bordeaux, in France. Since then, downy mildew spread very quickly throughout Europe: in 1879 it was already in Italy, in the Oltrepò Pavese, a very interesting and beautiful grape growing area located in Lombardy. It was feared that the disease might cause the end of European viticulture. Actually (see box page 26) pretty quickly a very effective product able to control its causal agent was found.

Therefore, on the one hand European viticulture indeed risked its end because of the arrival from the American continent of no less than three important pests within a few years and, on the other hand, the appearance of these three diseases, and the energies spent to control them, in a certain sense marked the beginning of crop protection. Even today, after almost 150 years, the management of powdery mildew and downy mildew requires several treatments per season.

You may wonder, how did it go on the American side? Some believe that the Americans' fondness for whisky is due to the impossibility of making wine in the New Continent. But that's another story.

It must however be said that on the coast of California, thanks to the mild climate, the European grapevine, introduced by the missionary fathers, who produced sacramental wine, did not suffer attacks of downy mildew. Nowadays California produces some of the finest wines in the world, which I really appreciate. To tell the truth, during my study periods in the USA, I was very well trained on American wines by two colleagues, who were eminent scientists, working on grapevine diseases (see box page 27). Among the best Californian winemakers, some are Italian (see box page 28).

Starting from Georgia

The importance of viticulture and oenology for Georgia is highlighted by the iconic figure of the local Christianity: Saint Nino, the nun from Cappadocia who preached Christianity in this region in the 4th century B.C., becoming one of the most venerating saints of the Georgian Orthodox Church. Her grapevine cross, with which the saint converted the king of Caucasian Iberia, is made of grapevine shoots. This plant became a symbol that we still find in the frescoes and bas-reliefs of the many monasteries in this area, places of worship but above all centres of protection and agricultural development of grapevine—true guardians of the biodiversity of the countryside and rural traditions, even when the Soviet Union exploited Georgia as the "cellar of the empire".

To get an idea of the number of native grape varieties of the area, just think that in the Georgian ampelography, edited by Ketskhoveli, Ramishvili and Tabidze in 1960, are identified and listed 524 indigenous grape varieties, compared to the 60 Italian ones and little less French.

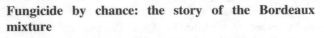

Going to depth

Fungicide by chance: the story of the Bordeaux mixture

I believe that even the less initiated in the field of agriculture have heard talking of Bordeaux mixture at least once in their life. This fungicide, based on copper sulphate and calcium hydroxide, was discovered in France by the botanist Pierre Marie Alexis Millardet in 1885, during a very serious epidemic of downy mildew on grapevine. It seems that Millardet made this discovery by chance, observing the effect of some practices used by the farmers of the famous grapevine growing region of Médoc, north of Bordeaux, to protect their vineyards, or better their grapes, from the robbers. To discourage thieves, growers used to apply a lime solution prepared in copper pots on the vine rows bordering the road. In 1882 and in the following years, downy mildew attacks on grapevine in Bordeaux region were very serious. Along country roads lined with vineyards, Millardet observed that the rows sprayed by growers with the lime solution had not been affected by the pathogen, while all other plants showed severe symptoms of downy mildew. This led him to make hypothesis on the effectiveness of lime and copper association against downy mildew. In order to confirm his hypothesis and well define the relationships between copper sulphate (the product with fungicidal action) and lime (capable of neutralizing the acidity of copper sulphate), Millardet had to conduct several trials with a mixture of copper sulphate, until he was able to perfect the mix with the most effective fungicidal action coupled with the minor toxicity. Sprayed on grapevine, this "Bordeaux mixture", so called from the region of the first observations, was able to protect plants from downy mildew. Thus modern crop protection was born. Within a few years, this fungicide became common in France, Italy and in all grapevine growing areas, where it is still well-known and used nowadays, with some variations to reduce its phytotoxicity. Other copper-based fungicides have been and are widely used, especially in organic farming, although in recent years, to avoid excessive accumulation of copper in the soil, maximum amounts of usage per season have been defined.

Learning about American wines

Two wonderful American colleagues introduced me to the American wines. In the late 1980s, during my stay at Cornell University, Roger Pearson, plant pathologists at the same University, introduced me to the then growing industry of New York State wines. Besides interacting in the lab, every other week we had very interesting wine sessions at his home. In the case of California wines, it was the late Doug Gubler, from U.C. Davis, who introduced me to the wonderful red Californian wines. Two very much renewed scientists, two wonderful friends, that I miss so much!

A former student to be proud of

Diego Barison, an initially very shy and provincial student, did follow in the early 2000s an International MS Degree in Sustainable Agriculture that I did develop, throughout an International Programme, at the University of Torino, in partnership with Universities in France and Norway. Day by day, Diego developed an international attitude, and, after spending one year of his *curriculum* abroad, he moved to work in France and later in California. Now, in California he owns two farms, dealing with the production of grapevine rootstock. More recently, he brought to USA the wonderful Italian meat. Diego, one of those students to be very proud of!

Coffee Rust in Ceylon: Why English People Drink Tea

Today Ceylon (now Sri Lanka) is worldwide known for its tea production, but few of us know that in 1870 this island was the main international coffee producer.

Let us go back in time to understand how a fungal pathogen was able to completely change a country's economy, affecting coffee and tea consumption all over the world.

Coffee is a tropical crop, native to Ethiopia. However, it was the Arabs who first used its seeds to prepare a new beverage that would later be highly appreciated, spreading everywhere. The scientific name of the plant, *Coffea arabica*, finds in fact its reason in the initial coffee production area.

Coffee is among the most profitable consumer goods (second only to oil) of the global economy. Imported into Europe for centuries, where it assured significant revenues for countries with colonial possessions in tropical areas, coffee has also taken on considerable importance in the colonies themselves—as they progressively gained independence—and has become a vital cash crop for many developing countries. The first places—the so called coffeehouses—in which costumers could taste this modern beverage date back to 1500s in Arabia, Egypt and Turkey (see box page 31). In 1600s coffee was a very popular beverage in Europe. After all, at that time, for hygienic reasons, you could drink water only if boiled and then coffee and tea were consumed abundantly. Tea was more popular because it was less expensive, while coffee was considered a more aristocratic beverage.

The Dutch were, in Europe, the first importers of coffee, which they brought in from their colonies in Ceylon, Java and Sumatra.

In Napoleon's time the Netherlands had to cede a large part of their colonies to Great Britain and in 1825 the British began to develop their coffee crops in Ceylon. At that time, every piece of land was used for the production of coffee and thousands of Indians went to Ceylon as labour.

By 1870, Ceylon had already become the first coffee producer in the world, but just that year a very dangerous pathogen, *Hemileia vastatrix*, agent of coffee rust, reached the island. This is a particularly dangerous pathogen as, after infecting the host, it reproduces and spreads very quickly, thanks to the release of millions of spores that, together, form the typical rust-coloured "spots" after which the disease is named. Ceylon coffee crops, which such dense plants and in presence of a humid hot climate, could not escape the violent rust attacks, with dramatic consequences.

To give an idea of the economic impact of the sudden appearance of this disease, it is enough to think that coffee production fell from 45 million kg in 1870 to 2.5 million in 1889. Within less than twenty years, most of the coffee estates collapsed and Ceylon had to switch its production to alternative crops not affected by this pathogen.

A few years after the potato blight epidemic in Ireland, history was repeating itself on a different crop, in another continent, with a different disease.

Coffee rust epidemic had devastating consequences on Ceylon's economy. Again, the sudden appearance of a pathogen led a country to starvation, this time not because coffee was a staple food for the population (as was the case of potato in Ireland), but because the whole agriculture of that country had converted to coffee production, for economic reasons imposed by colonial powers.

Thousands of Indian workers returned home and the remaining farmers began to convert their lands to tea production, while coffee crops moved to the western hemisphere, where today Brazil and Colombia are the main producers.

The switch to tea production in Ceylon was not immediate: farmers had to look for the most suitable varieties and marketplace, therefore it was really a hard time for the country. Fortunately, the cultivation of tea was successful, no new parasites affected this crop and, in the meantime, the development of the first fungicides had provided means to control diseases, if ever needed.

That is why today we identify Sri Lanka (former Ceylon) as a tea-producing country, and why British converted to this beverage.

As in the case of the late blight of potato, it is interesting to see what happened later to the coffee rust fungus. After the devastations caused in Ceylon, in the new coffee cultivation areas it did not cause problems for a long time, thanks to the adoption of "quarantine" measures (see page 201) similar to those used to avoid the spread of human and animal diseases. These measures, taken to prevent the introduction of the pathogen in a new cultivation area through the strict control of plant material, proved to be effective for more than one hundred years. It was confidently assumed that coffee rust could not cross the cordon sanitaire of the Atlantic. That was wrong. No one can say how rust came to the Americas. It might have arrived in shipments of other plants, living or dried. It might have clung to the shoes or clothes of travelers. It is even possible that rust crossed the planet on high-altitude winds, the route that another plant disease, wheat-stem rust, has used to spread between continents. Coffee rust moved without detection, and then, in 1970, its telltale spots and spore-laden dust appeared on coffee plants in Brazil. It spread quickly west and then north: to Peru, Ecuador, Colombia, and then up through Central America. The disease was fierce, but when it appeared, repeated application of fungicides and careful management of plants kept it in check. For a while, since recently new severe epidemics occurred (see box page 32).

Every morning, when we taste our first coffee, let's think that around the world, 100 million people draw dignity and income from coffee, one of the world's most traded agricultural products. A product now again at risk in many producing areas. Once again, history repeats itself!

Coffeehouses, the Starbucks of the past

In 17th century, coffeehouses were very popular in England: they were more than 3,000! Each place had its own specific customer, based on the profession or on political trends. Also the Venetians, thanks to their trade, were among the first to taste the new beverage, appreciating it very much: at the end of 1600s there were more than a hundred "Botteghe del Caffè" in Venice, and one of the most beautiful comedies by Carlo Goldoni, written at the end of 1700s, is just titled "La Bottega del Caffè" (*The coffeehouse*).

Coffee rust, again a problem for small farms

In 2008, rust flared up in Colombia as devastatingly as it had in Asia 150 years earlier, and by 2012 it had moved into Central America. As it had in Ceylon, it wiped out entire farms. In tens of thousands of small farms across Central and South America, coffee plants are stumbling under the assault of rust. In some areas, more than half of the acreage devoted to coffee has ceased producing. From 2012 to 2017, rust caused more than $3 billion in damage and lost profits and forced almost 2 million farmers off their land.

Coffee is a lifeline for small farmers in areas too thin-soiled or forested or steep to grow much else. Climate change—more heat, more intense rain, higher persistent humidity—created, very probably, conditions that made coffee farms more hospitable hosts. In 2012, temperatures across Central America were higher than average; rainfall was erratic and drenching. Together, those phenomena allowed the rust to cycle more rapidly through its reproductive process: infecting the leaves of a plant, generating spores, releasing the spores, and finding a new plant on which to grow.

Bitter Rice

In 1942, while everyone's attention was focused on World War II, India (which was then an English colony) was threatened with attack by Japan.

At that time, as it is today, the Indian diet was rice-based. In the north-west of Bengala, up to three rice crops could be cultivated per year: a main and two smaller ones. The main crop benefited greatly from the rains of the monsoon season. Until 1942 in the Bengal region the fungus *Helminthosporium oryzae* had caused very limited damage (see box page 34), but unfortunately, in 1942 the particular climatic conditions, with an extension of the monsoon, caused more abundant rains than usual, that favoured the sporulation of the pathogen from the smaller cultivations to the main one. Yield losses reached 70–90%. This substantial decline in production pushed the rice prices higher at the beginning of 1943.

The Indian government, focused on the war, did not address the problem with the necessary promptness and attention. Neighboring Myanmar (then Burma) was occupied by Japan, which prevented Indians from accessing their traditional food.

Many Indians, no longer able to obtain their main source of nourishment, starved to death or were victims of other famine-related diseases. More than two millions people lost their lives in 1943: a tragedy similar to that caused by potato blight in Ireland. Farmers began to leave their lands, moving to the cities in search of work.

This Bengal one was the most recent of the major famines caused by a disease. Brown spot disease, whose epidemic afflicted the rice crops of the country, is still present nowadays but can be managed by various means, including fungicides and use of resistant varieties.

Rice remains a main crop in many areas of the world and millions of farmers still struggle in protecting it from many pests (see box page 34).

M. L. Gullino, *Spores*,
https://doi.org/10.1007/978-3-030-69995-6_6

Why an already present pathogen may suddenly become dangerous?

It is very impressive to see how a plant pathogen already present and quite well tolerated for several years in the Bengal region suddenly became as harmful as to cause millions of victims. Certainly, the exceptional climatic conditions resulted favourable to plant infection, thus referring to the already mentioned "triangle" model we can say that this time it was the environment to trigger the epidemic.

A movie on the hard work of rice weeders

Riso amaro (Bitter rice) is the title of a very famous Italian movie directed by Giuseppe De Santis in 1949, with Silvana Mangano, as main actress, playing the role of a rice weeder. Such a movie permitted people to understand the harsh life of rice workers, before the development of herbicides. Later on, the use of herbicides in rice fields, has gone throughout much discussion, also for some collateral effects (i.e. their drift into close-by fields, such as vineyards). Difficult, anyway, to consider going back to the time of hand-weeding of rice.

Mycotoxins: How Many Deaths in Africa?

A number of plant pathogenic fungi cause not only direct but also indirect damage, producing secondary metabolites called mycotoxins that can be extremely detrimental to both animals and humans. These substances became infamous in the early 1960s, when 100,000 turkeys died in England from feed contaminated with a particular type of mycotoxin, aflatoxins. It was referred to as the "turkey X disease".

Aflatoxins are the mycotoxins known for the longest time, the most dangerous for their confirmed carcinogenic action on liver. Aflatoxins are now considered to be responsible for the high rate of hepatic cancer reported in Africa. *Aspergillus flavus* and *Aspergillus parasiticus* are the main aflatoxins producers. The former is more adapted to temperate climates, while the latter finds in tropical and sub-tropical areas the most suitable environmental conditions (high temperatures and relative humidity) for its development.

There is therefore an important myth to dispel: aflatoxins are not just a third countries concern. Certainly in developing countries, climatic conditions conducive for the development of pathogenic fungi combine with non-optimal grain and food storage conditions. However, in recent years also in our temperate areas environmental conditions have been favourable to the emergence of *Aspergillus flavus*, with consequent contamination in corn and other agricultural products.

The high incidence of liver cancer observed in Africa is now explained by the prolonged consumption of aflatoxin-contaminated foodstuffs, even at low levels, which causes severe chronic diseases.

In addition to cereals and their derivatives, dried fruits, herbs, milk and tea are some of the many products that can be contaminated by aflatoxins. Today aflatoxins, as well as other mycotoxins described later, are subject to specific regulation in many countries and limits of aflatoxins have been set for several products. Food safety legislation on aflatoxins has changed over time, becoming more and more stringent, especially for baby food.

The possible contamination by mycotoxins of foodstuffs produced in third countries greatly complicates their agricultural economy: in fact, importing

© The Author(s), under exclusive license to Springer Nature Switzerland AG 2021
M. L. Gullino, *Spores*,
https://doi.org/10.1007/978-3-030-69995-6_7

countries carry out regular checks on imported goods and third countries often fail to meet the required standards, exceeding the limits allowed by the importer.

In recent years also in Italy the weather conditions did favour the development of *Aspergillus flavus*, with important contamination on corn. Aflatoxin-contaminated feed consumption has led to contamination of milk with aflatoxin M, a mycotoxin we find in milk produced by cows consuming aflatoxin-contaminated feed. As a consequence, thousands of gallons of contaminated milk had to be removed from the market. Aflatoxins are certainly the most important, investigated and dangerous mycotoxins, but they are not, unfortunately, the only ones.

Today the problem of mycotoxin-contaminated foodstuffs is real and current: flours, drinks, feed, spices can be contaminated by different mycotoxins, which threaten animal health especially with chronic diseases. Mycotoxins are generally very stable and difficult to eradicate by physical, chemical and/or biological means.

Crazy horse not only in Paris

When you say Crazy Horse, the older people think of the famous Parisian club and Rosa Fumetto. Plant pathologists, on the contrary, think of a mycotoxin-induced disease that mainly affects horses, due to the ingestion of fumonisins-contaminated feed. Fumonisins are a group of mycotoxins produced by *Fusarium*, which cause a lesion in the brain of horses inducing life-threatening symptoms, such as blindness and a state of serious restlessness.

Plant Diseases in Industrialized Countries

What we have seen so far perhaps gives you the wrong impression that serious epidemics only occur in developing countries or are something of the past. But it's not like that, and the story I'm about to tell you will confirm it quite effectively.

United States of America, 1970: we are in the famous "corn belt", a wide intensive corn cultivation area. At that time, American growers all cultivated some selected hybrids of corn, containing a cytoplasmic factor (so-called T factor, capable of maintaining male-sterility). Although very useful for hybridization work, this factor nevertheless caused a susceptibility towards a particular fungal race (T) of *Heterosporium maydis* (now called *Bipolaris maydis*), capable of producing a very selective toxin, T-toxin, active on corn.

That season was a disaster: the enormous selective pressure exerted by the cultivation of a few corn hybrids, containing the T-factor, over millions of acres, favoured the spread of the T-race pathogen. The damage was huge, as the disease affected the 80% of the cultivation area.

Obviously, being an epidemic of an annual species such as corn in an industrialized country in the 20th century, it caused mainly economic losses. Already the following year, the corn hybrids previously used, characterized by extreme susceptibility to the pathogen, were replaced. Crop failures, however, resulted in severe losses for corn growers.

The main culprit of this catastrophe was the cultivation on wide surfaces of very few hybrids with very similar genetic profiles. Genetic uniformity potentially exacerbates disease epidemics, a risk run every time that, on intensive cultivation areas, few crop varieties are grown on large surfaces. A typical risk of the most advanced agricultural systems.

Tulips with Fever

Few of us know that the first great economic crisis—a real financial crash—was unleashed by tulips in the Dutch Republic of late 1600s. Who would have said that the fondness for this flower, made popular by the Ottoman Turks, could have led to such speculative mania? Let's try to retrace this curious and fascinating story.

Around 1550, a Flemish diplomat, Ogier Ghiselin de Busbecq, introduced tulips to Europe after discovering them in Turkey, where they were wild species. Tulips immediately became fashionable (see box page 41) and sparked the interest of some farmers who began to grow them. The most requested were the mottled ones.

It is now well-known that some viruses —microscopic and very insidious agents —can cause in the infected flowers and plants a striped multicolored pattern which is in fact a clear disease symptom. Bulb plants —and tulips in particular—are among the more affected species.

At that time, mottled tulips became very popular for their pattern variation (which was not yet associated to viral infection) and were portrayed in several masterpieces by painters of the Dutch and Flemish Schools. This fashion unleashed a general botanical-financial excitement in the Dutch Republic (see box page 42).

Tulip bulbs were used as currency and exchanged for pigs, sheep, wine, butter, clothing and furniture. Mottled tulips reached incredible prices and the rarest and most valuable hybrids reached so insane quotations as to be worth more than a house. It was a real "tulip mania". People's interest in this flower no longer depended on its beauty, but on the desire to yield economic benefits from it.

Tulip fever affected everyone: simple peasants, who considered them a source of infinitely higher income than other agricultural products, and greedy merchants who aimed only at profit.

Born as a harmless and extravagant fashion in tight circles, tulip mania was then exacerbated by speculators who pushed the value of tulip market beyond any logic. The highest price ever recorded for a tulip bulb in 1637, when the bubble had burst, corresponds to 5,200 florins.

© The Author(s), under exclusive license to Springer Nature Switzerland AG 2021
M. L. Gullino, *Spores*,
https://doi.org/10.1007/978-3-030-69995-6_9

The average annual earnings of a wealthy merchant around 1630 corresponded to 3,000 florins and in 1642 Rembrandt's *Night Watch* was purchased for 1,600 florins: this can give an idea of the insanity achieved.

Primitive financial contracts were introduced and used, which allowed the trade, in real money, of bulbs that in fact did not exist, as they had not yet been planted. The expectations of resale on advantageous terms became the determining factor of the exchange value that set the tulips' price *ex ante*, without any real counterpart.

It resulted in a collective delirium that can be summarized as follows: everybody buys tulips, because everybody else wants to buy tulips, because everybody expects everyone to want to buy tulips, and so tomorrow I can sell my tulip at a higher price than the one I bought today.

As a result, the financial bubble swelled to overflowing and in February 1637 bursted, bankrupting all those who had speculated on the tulip trade. As Monsieur de Blanville wrote in 1743 in his book *Travels to Holland*, "They were possessed by such furious passion for those flowers or, to call it by its name, by this itchy desire that they often offered three thousand crowns for a tulip that would satisfy their fantasies: a virus that ruined many rich families".

The tulip fever was described by Deborah Moggach in 1999 in her novel *Tulip fever* and became a movie in 2017, directed by Justin Chadwick. Something that plant pathologists should read and watch!

In addition to the phenomenon above described, which led thousands of Dutch families to bankrupt, two episodes of collective hysteria, equally curious, though with less devastating consequences, occurred among the Turks of the Ottoman Empire.

In Turkey, tulip mania had a more artistic and cultural component. But what was so special about the tulip? It must be said that tulips of that time were really stunning flowers, much more beautiful than the varieties relatively insignificant commercialized today. Their petals showed intricate and original patterns, with vivid colours, which made them highly enticing.

In Islamic gardens, tulip was regarded as the holiest and most precious of the flowers, thus Turkish fondness for this blossom went well beyond a simply admiration for its delighting beauty. Tulip was considered by the Ottomans the flower of God, because in Arabic culture the letters that make up the word *lale*, 'tulip' in Turkish, are the same that form the name 'Allah'. It was the symbol of the Ottomans' humility before God, as at the peak of flowering, it bows its head.

In the Ottoman illustrations of the Garden of Eden, tulips are often present, in full bloom, under the fruits of the Tree of knowledge of good and evil. The afterlife imagined by the Turks, ready to sacrifice themselves in battle, was a garden full of blooming tulips. That's why these flowers accompanied the Turks in their westward advance.

Something different occurred some years ago in the United States, yet the two examples prove how history is repeated over time (see box page 42). Tulip trade is sometimes used nowadays as a cover for illegal drug trafficking (see box page 43).

Even today, Holland is one of the main producers of bulbs in general and tulips in particular, although, as we will see later in these pages, bulbs are no longer produced in the Netherlands but by Dutch companies in third countries while the Netherlands simply market them worldwide. Since then, however, tulips are sold at wholesale prices and the famous Royal FloraHolland is one of the largest flower auction in the world (see box page 43).

Curiosity

Tulips as ornaments

Around 1610 the flower mania became more and more widespread in Paris, where the most prominent aristocrats would offer their ladies the rarest species. The most beautiful varieties were as appreciated as diamonds. The refined elegance of tulips led them supplant roses. At the wedding of Louis XIII, in 1615, the finest ladies of the aristocracy wore tulips in the neckline of their gorgeous robes.

Bubbles through history

Tulip mania is really something to think about. It represents, in fact, the first financial bubble of history, which heralds and anticipates the subsequent ones: tulips in 1600s as the subprime mortgages in 2000s. In both cases, the trigger was a collective delirium, an unstoppable social contagion that led the system to collapse. In the Dutch Republic it was a flower to unleash human insanity and greed.

Such irrational conduct has had many faces and various ways of expressing itself in history, though its first reification was in a mottled tulip affected by a virus. Its amazingly colourful petals made it so alluring to people as to bring them to insanity: they would suddenly squander their wealth to get nothing but scentless and perishable flowers.

The new economy crash, the housing bubble and the resulting subprime mortgage crisis, whose consequences are still evident, have been just other examples of that irrational behaviour experienced with tulip mania.

As already seen and as we will see again with other examples, history repeats itself. If the recent economic crisis, starting in the US at the end of 2006, has reached such proportions as to lead millions of people to bankruptcy and destabilize the world economy, it is probably because in 400 years we have not yet equipped ourselves with the right antibodies—some researchers say.

Drugged tulips

Marijuana and tulips. Flowers grown in Dutch greenhouses, loaded on a lorry and brought to Rome together
with hundreds of "space cakes" and bags of "grass".
Four trips a month, over 50–60 kilos of drugs at a time.
In 2004 a real criminal organization dealing with drug
smuggling behind tulips' trade was found out by police:
the network involved the students of a dozen Roman
schools and the faculty of Psychology of the Sapienza
University of Rome.

Going to depth

The Royal FloraHolland

The floral auction in Holland, with its famous Florists
Clock, is first on the list for wholesale buyers from all
over the world and has always been a unique
experience for the neophyte. Large batches consisting
of hundreds of thousands of flowers of the same
species, come to auction from all over the planet to be
then redistributed into smaller batches, along with
different species, so that each florist, exporter or
wholesaler can always receive the right amount of
flowers and plants for their needs. The final recipients
are exporters and chain stores, but also florists and
market vendors, who sell single bunches of flowers. At
FloraHolland headquarters in Aalsmeer, buyers can
choose products with an extraordinarily large assort-
ment from countries such as Kenya, Ethiopia, Israel
and South America. An average number of almost
116,000 clock transactions per day, an annual turnover
of 4 billion euros, 6,500 suppliers, 5,600 traders and a
trade of over twelve billion flowers and over half a
million plants per year: these are the impressive auction
data.

Why Kissing on New Year's Eve Under Mistletoe Brings Good Luck?

Mistletoe (*Viscum album*) is an evergreen epiphitic plant parasite of several species of trees, such as poplars, oak, lindens, elms and pines, from which it draws water and nutrients. It is able to perform photosynthesis (hence the green colour), but needs other plants' contribution for its growth.

Largely known and entwined with Christmas season, mistletoe is considered the first plant pathogen recognized as such. Early around 1200 A.D., in fact, Albertus Magnus described mistletoe as a parasite of plants and at that time go back real advice on how to control its development with cultural practices such as pruning.

Practically mistletoe is a plant that, as a parasite, lives at the expense of other plants, which thus will have a reduced development and be more subject to the effects of adverse climatic conditions (e.g. winds and thunderstorms), producing wood of lower quality.

The fact it maintains its bright colour even in winter, when the host plant loses its leaves, inspired Greeks with traditions and myths. Depending on the place and on the time, mistletoe was attributed aphrodisiac properties, able to aid conception, or antitoxic properties.

The first mention of mistletoe associated with particular powers was by Pliny the Elder, a Roman author and philosopher writing about nature. Pliny scoffed the druids of 1st century B.C. who were persuaded that an infusion of mistletoe would ensure fertility to all sterile animals.

It was also thought that the golden branch that allowed Aeneas to descend into Hades in the *Aeneid* was just the mistletoe, at that time considered a symbol of vitality by many populations for its evergreen nature (see box page 46). Mistletoe was hung on the ceilings or doors to keep away witches and evil spirits; the Romans used it to decorate houses in winter, as an homage to gods. In the Nordic countries mistletoe was dedicated to Frigg, the goddess of love and marriage, and in Scandinavia it was considered a plant of peace, under which peace treaties were signed and marriage agreements were made (see box page 47).

It seems that the tradition to hang mistletoe as a Christmas decoration started in England. In a novel of 1820, W. Irving consider mistletoe among Christmas

M. L. Gullino, *Spores*,
https://doi.org/10.1007/978-3-030-69995-6_10

ornaments and it was Charles Dickens in 1836 who first mentioned a mistletoe-blessed kiss. In a scene of the *Pickwick Circle*, in fact, he writes that the younger ladies, finding themselves under the mistletoe, "screamed and struggled, and ran into corners, and threatened and remonstrated, and did everything but leave the room, until… they all at once found it useless to resist any longer, and submitted to be kissed with a good grace". In this case, mistletoe was seen as a lucky charm for couple kissing under the plant. And so, by combining myths and traditions, the belief was established that kissing on New Year's Eve under a branch of mistletoe brings good luck. Why not believe it?

Some say that at each kiss the men had to take off one of the berries from the branch: the kissing would have ended up once finished the berries. Let us remember it on next Christmas, together with the story of this humble parasitic plant.

The golden branch

The English scholar James George Frazer in the 20th century dedicated his colossal study (12 volumes) *The Golden Bough* to popular traditions such as, in the ancient cult of Diana, the custom of cutting a branch of a particular tree in the woods, before slaying the priest, something that would be connected to the origin of royalty. In conceiving his work, Frazer was highly inspired by the homonymous painting of Turner, depicting the episode from the *Aeneid* by Virgil where Aeneas descends into Hades by means of the golden branch.

The myth of Frigg

The myth tells the story of Frigg and her son Baldr. According to the legend, worried that something might happen to her son, Frigg gathered every object on earth (plants, animals, stones etc.) and made them vow never to hurt Baldr. Frigg, however, neglected to consider the unassuming mistletoe, which then Loki, God of chaos and deception, used as an arrow to kill Baldr.

According to another version of the myth, it was Frigg's tears that turned into mistletoe berries, so that she decided to make this plant a symbol of love. Frigg managed to bring her son back to life under a plant of mistletoe, then, delighted, she promised a kiss to all those who passed beneath it. At the same time, in order to punish mistletoe for having wounded her son to death, she condemned it to become a parasite unable to cause pain or death, and decreed that anyone who passed beneath the mistletoe would enjoy peace and love.

Changing Landscapes

Some serious diseases can affect not only agricultural crops but plants that have a great importance in characterizing and beautifying the landscape.

The earliest written works on ornamental plant diseases date back to the end of 1700s, when Monsieur Liège wrote *Le jardinier solitaire*, while at the beginning of 1800s the Emilian botanist Filippo Re devoted some attention to devastating infections of ornamental plants such as root rot diseases, which were treated at that time by eradicating parts of the sick roots and replacing the soil around the affected plants. Other fungal diseases were mentioned, such as powdery mildew, well recognizable for the presence of a white coating on the leaves.

Many examples could be given, all very interesting, but this book is not a treatise on plant pathology. So, among the various cases, I've chosen to describe the two that, for virulence, peculiarity and geographical proximity have been most relevant in the last century.

Let's start with the story of a serious elm disease, best known as Dutch elm disease. The causal organism, once called *Graphium ulmi* and more recently renamed *Ophiostoma ulmi*, affects all species of elm, European and American.

In Italy, the mountain elm tree grows mainly in the alpine zones and on the central-northern Apennines, whereas the rural elm was used as woody and ornamental species. The rural elm was widespread in parks and gardens, and used for tree-lined streets. In many wine-growing areas it served as a support for vine rows, so much so that it was commonly spoken of "married to elm vine" (see box page 52). Moreover, its wood was used for making furniture, floors etc.

The elm was therefore, in the early 1900s, a well-integrated species in the Italian landscape. Dutch elm disease was first reported in the Netherlands in 1922, but it is believed that it was already present there for several years as it was in Belgium and Northern France, from where it quickly spread to Germany, England and Central Europe. In Italy the disease arrived in 1929 near Modena and was reported in 1930. In the same year, it reached the state of Ohio, in the United States of America, and again it spread rapidly to the west and Canada. Its common name, "Dutch elm disease" (DED), reminds therefore of its geographical origin.

© The Author(s), under exclusive license to Springer Nature Switzerland AG 2021   49
M. L. Gullino, *Spores*,
https://doi.org/10.1007/978-3-030-69995-6_11

The first symptoms of the disease appear at the beginning of summer and consist in yellowing, withering and desiccation of leaves, followed by desiccation of branches that often, at the apex, bend to hook. Within one or two years the affected plants wilt and die. The causal agent of DED, the aforementioned *Ophiostoma ulmi*, as well as spreading from affected plants to healthy ones by taking advantage of the continuity of tissues that is created at roots' level (radical anastomosis), exploits some bark beetles that spread the spores of the pathogen by digging tunnels in the woody tissues of still healthy plants. In other words, these small insects are formidable pathogen vectors.

In the years 1920–1940 Dutch elm disease spread as well as in Europe and in many areas of North America, also in Asia, causing considerable damages on the economy, environment and landscape of urban and rural areas.

To this first epidemic wave, to which the elms partially managed to survive, after what was supposed to be a cooling-off period, in the years from 1960 a second wave followed, due to a new strain of the pathogen, much more aggressive, which caused even more serious damage, wiping out the left elms. Canadian wood was blamed for this second DED wave, as it was said it had introduced the pathogen into United States.

Over the following 15 years almost all European elms were affected by graphiosis. The impact was this time enormous, both on the economy and the landscape, but also from the emotional point of view.

In the 1970s a postcard saying "Our generation is the first to see a man on the moon and the last to see an elm on earth!" became popular in the United States. Also in Italy the ornamental plants of the historical parks were highly damaged, also because the elm was among the most commonly used plants in the garden design from 1600s to 1900s.

In view of the importance of the affected species, the severity of the disease and the virulence of the pathogen, efforts have been made to develop management methods based on the eradication of diseased or dead plants, as they might serve as a source of inoculum and vectors, the small elm bark beetles.

The affected trees need to be felled and the infected wood burnt or buried. In order to control Dutch elm disease, insecticides have also been used against elm bark beetles and, in order to manage the fungal pathogens, "endotherapy" measures have been developed, which are injections of systemic fungicides able to move within the woody vessels invaded by the pathogen. The best results for elm recovery, however, have been provided by genetic improvement. Both Dutch and American governments have invested several resources for the management of DED through tolerant elm genotype selection and breeding programmes.

Today, after decades, more resistant varieties have been successfully introduced, so that elm trees are slowly returning to American boulevards, gardens and parks, though in the countryside and in the woods, the elm seems destined to survive only as a bush.

Another very interesting example is cypress canker. Once again, we are faced with a kind of high aesthetic and landscape value, common especially in Tuscany, region of central Italy, where countryside and roads are often adorned with cypress trees. We find this typical landscape in many Renaissance paintings, and cypress trees are honoured by Giosuè Carducci in his poem *Davanti San Guido* (In front of San Guido): "the cypress trees stand straight and true from Bòlgheri to San Guido in double rows…". To the cypress as a symbol of death brings back the incipit of *Dei Sepolcri* (The graves) by Ugo Foscolo:

Under the shadows of the cypress trees,
within the urns wetted by loving tears
can the slumber of death be less profound?

Also this beautiful tree, always associated to the Mediterranean landscape, has suffered severe infections, as in the mid-1900s a fungus called *Coryneum cardinale* (later renamed *Seiridium cardinale*) attacked Italian cypresses causing a life-threatening canker that affects their twigs and branches.

In fact, this disease was first reported in California, in 1928; a few years later, in 1933 it was reported in New Zealand and in 1944 in France. It arrived in Italy in 1951, attacking young cypress trees in the monumental Cascine Park in Florence.

It is believed that the pathogen came from France, through the import of young cypresses, already infected in latent form by the pathogen. Cypress canker quickly spread to Tuscany, Umbria and all regions where cypress was grown. The same happened in all the Mediterranean countries.

The harmful impact of this cypress parasite is visible everywhere. As already mentioned, this plant had an important role for landscapes, nurseries and forests. The varieties of *Cupressus sempervirens* (the Italian cypress) were also used for creating pure or mixed cypress woods, which often had the task of enhancing land with poor soil characteristics.

Besides the virulence of the fungal pathogen, the severity of this disease is also due to the artificial introduction of the cypress into not properly suitable areas. This species, in fact, outside its home range, is often subject to cold damage whose symptoms are more or less evident wounds that are for the pathogen the preferential way of entry into the tissues of the plant. Once penetrated into the plant, the fungus develops rapidly causing necrosis of the tissues with which it comes into contact. The plant opposes a strong reaction that is manifested with emissions of an abundant quantity of resin and with the production of a barrier of cork cells which however are overcome by the fungus when the plant goes into vegetative rest.

When grown in their home range, cypresses do not go into vegetative rest and are thus able to effectively counteract the colonization of tissues by the pathogen. When the fungus has colonized the entire axis of the branch, the top withers and the infection proceeds downwards.

Thanks to national and European funding, the genetic improvement carried out by researchers of the National Research Council of Florence led to obtain canker-resistant clones that are currently on the market and let to revive the cypress-lined avenues described by Carducci.

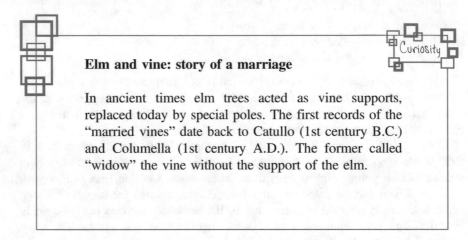

Elm and vine: story of a marriage

In ancient times elm trees acted as vine supports, replaced today by special poles. The first records of the "married vines" date back to Catullo (1st century B.C.) and Columella (1st century A.D.). The former called "widow" the vine without the support of the elm.

Plane Trees Die by Canker

Also the plane tree, another arboreal species much used for ornamental purposes (see box page 54), has suffered a serious threat of extinction due to the import from United States of a fungus, called *Ceratocystis platani*, which causes a lethal disease in all cases, known as "canker stain of plane" or "canker of sycamore". This disease, first appeared in the United States in 1925, spread in most Atlantic seabord states until the World War II. The war itself was the cause of its introduction in Europe, through the American ammo crates built with the infected plane tree wood and discharged in all the main ports of Europe. The outbreaks of the infection were found in Naples, in Marseille and in Barcelona. In Italy the disease was reported in 1971–1973 in Caserta (where serious losses occurred in the arboreal heritage of the Royal Palace)—although perhaps in this area it was already present for many years —and in 1974 in Marseille, in Barcelona, in Belgium and in Turkey.

Then it spread rapidly throughout the Mediterranean and northern Europe, thanks also, as already seen in other cases, to the increase and speed of transport and trade. The damage was very serious: Marseille lost in few years 20,000 plane trees.

The most visible external symptom is the sudden disruption of water transport within the infected plant, with consequent yellowing and desiccation of leaves. The phenomenon may affect the entire plant or part of it and is due to the occlusion of the vascular tissues by the fungus and the consequent impossibility for the plant to get the raw sap to the leaves.

The name of the disease, "canker", comes from the typical necrosis of inner bark, while the word "stain" refers to the discolouration of sapwood with streaks going from reddish-brown to bluish-black. Canker stain of plane tree not only affect and kill very valuable specimens, with damages to the landscape and economic losses for the local administrations that manage parks, gardens and trees, but can also cause the falling of big branches that might strike things or people passing by.

When plane trees are very close to each other, for example in the case of tree-lined avenues, the infection spreads very quickly from the affected plants to the healthy ones, due to their proximity at root level and to the many ways of

© The Author(s), under exclusive license to Springer Nature Switzerland AG 2021   53
M. L. Gullino, *Spores*,
https://doi.org/10.1007/978-3-030-69995-6_12

penetration and diffusion the pathogen can use, such as pruning wounds, which are a common consequence of the interventions carried out almost every year to control the development of the foliage, and sawdust produced by the chainsaws used for pruning or to uproot infected and dead plants. Moreover, the sawdust containing the spores and the mycelium of the pathogen adheres very easily to the tires of the cars that in the cities are often parked under the plane trees, and from there it reaches healthy plants.

To understand how dangerous is the canker stain of plane tree, suffice it to say that in the 1970s, within 7–8 years in the Royal Palace of Caserta alone, 900 bicentenary plane trees had to be demolished and that the magnificent tree of the square in front of the Pontifical Basilica of Saint Anthony of Padua suffered a real scourge.

This is the reason of the numerous gaps which are often noticed in tree-lined avenues of many cities, caused by the death of large plane tree specimens and filled at times with young plants belonging to other species.

The severity of this disease made it essential, in 1987, a Ministerial Decree of compulsory control forced to burn or remove all the affected plants (see box page 55). In fact, prevention made it possible to greatly reduce the spread of the pest: the city of Turin is therefore a positive example in this sense.

Famous plane trees

According to a legend, Hippocrates of Kos used to teach his pupils the art of medicine, of which he is considered the founding father, under the foliage of a plane tree. According to other sources, also Paul of Tarsus would preach under a plane tree.

Today, in the town of Kos, in the "Square of the Platane", just in front of the Town Castle, there is a plane tree of about 500 years, which is said to be the descendant of that famous specimen which stood in the same place at the time of Hippocrates.

It is mandatory to fight!

In the case of very serious diseases, caused by very aggressive pathogens, which spread very quickly, it has at times been necessary to issue mandatory control decrees forcing to remove and destroy the affected plants. Such interventions have made it possible to control the infection outbreaks and limit the severity of the pathogen attacks.

In the case of canker stain of plane, the decree requires the infected plant and the two lateral ones to which the pathogen might have already spread by means of roots' contact to be felled; infected wood must be transported on flatbed tarpaulin trucks to the appropriate disposal areas; pruning must be carried out according to precise rules; new plantings in the felling sites are forbidden; in the new plantings, trees have to keep a distance of at least 40 foot from each other, in order to avoid contact between the roots of nearby plants.

Destroyed Forests in California

The "story" of *Phytophthora ramorum* is another example of the economic and environmental impact of introducing an exotic pathogen on forest plants.

At the end of the last century, this species of *Phytophthora* has been considered responsible of a serious decay of the oak forests ("sudden oak death") in California and in Oregon (United States of America) (see box page 58). Later on, the pathogen was found in Washington State and in Canada, on rhododendron plants grown in nurseries. In Europe, it has been reported, especially on rhododendron and viburnum, in the nurseries of many countries (Belgium, Denmark, France, Germany, Ireland, Netherlands, Norway, Spain, Sweden and Great Britain). Finally, the pathogen has been detected also in Northern Italy, but only on pot plants coming from abroad. In the American oak forests, the pathogen, due to its ability to attack different species, would seem to threaten also the biodiversity of the undergrowth flora. In Europe, *Phytophthora ramorum* is considered a pathogen that can seriously compromise the quality of ornamental plants, such as *Viburnum* and *Rhododendron*. In America, the strategy of eradication is carried on to fight *P. ramorum*.

In Italy, it seems that the eradication of the first outbreak, carried out with the destruction of infected imported rhododendron and azalea plant, was successful. In the year following the report, in fact, there were no subsequent reports of the presence of the pathogen in the area of Northern Italy where these ornamental species are very popular. With regard to this pathogen, however, just because of its ability to attack many ornamental species cultivated in pot and exported to distant countries, it is necessary to pay high attention.

M. L. Gullino, *Spores*,
https://doi.org/10.1007/978-3-030-69995-6_13

Good thing there's Matteo

One of the most famous and committed specialists on *Phytophthora ramorum* is the Italian Matteo Garbelotto, professor at the University of Berkeley, California. Matteo is very fond of his original country and every year he tries to return to Venice, his city, not to miss the Feast of the Most Holy Redeemer, an event held the third Sunday of July. Matteo also draws to this feast his forest pathologists friends from Turin, to which he is bound by an intense scientific collaboration and a deep friendship. Matteo has studied aspects of biology and epidemiology of *Phytophthora ramorum* in order to better understand its ways of spread, developing diagnosis methods and prevention strategies. Since 2020 he is also the Editor in chief of the Journal of Plant Pathology, which is the scientific journal of the Italian Society of Plant Pathology: one more signal of Matteo's link to Italy!

Climate Change and Plant Diseases

Climate change is a natural phenomenon, an environmental peculiarity driven by fluctuations of solar energy, volcanic eruptions, variations in Earth's orbit and ocean-atmosphere interaction. Human impact on the environment forced this variability by altering the mean values and the intervals within which atmospheric phenomena occur, inducing the so-called "climate change", whose effects are evident in the variation of atmospheric composition and in thermos-pluviometric regimes, in terms of intensity and frequency.

Towards the end of 1900s, as a result of more important climatic variations than in past decades, the researchers focused their attention on climate change, its possible causes and its potential effects. In addition, since the mid-1970s it has been noted that, despite advances in agricultural technologies (new, more productive cultivars, innovative disease management and fertilization strategies), climate continues to cause significant yield losses. These considerations have highlighted the need to understand how climate change can affect agriculture.

As regards economic aspects, variations in daily and monthly values of variables such as temperature and precipitation seem to have a marginal impact on plant production. Serious effects, on the other hand, might occur if the variations concern the extreme values—in particular, qualitative and quantitative impacts and geographical shifts of the cultivation areas are expected.

Fluctuations in climatic trend together with the alteration of the chemical composition of the atmosphere may also cause changes in the morphology and physiology of plants. For example, the increase in atmospheric CO_2 concentration can have fertilizing effects on crops (see box page 60). In fact, the intensification of photosynthesis and the improvement of efficiency in the use of water, found in environments enriched with CO_2, allow an increase in productivity. This, nonetheless, might be limited by simultaneous variations in temperature and precipitation, which not only can be a functional disadvantage to the crop, but also can make it more susceptible to pathogens and pests. The extent of plant growth changes in CO_2-enriched atmosphere conditions therefore also depends on other factors—climatic and non-climatic—such as the impact of possible diseases and host resistance to them.

M. L. Gullino, *Spores*,
https://doi.org/10.1007/978-3-030-69995-6_14

59

Potential impacts of climate change either are simulated with experimental trials carried out in a protected environment (phytotrons, growth chambers, etc.) (see box page 61) or in field, or are estimated through the use of climate simulation models combined with growth models.

Studies so far have shown that climate change can lead to variations in agricultural production due to direct effects on crop physiology and indirect effects through changes in the nutrient cycle, pest-crop interaction and emergence of new pathogens and pests.

Plant pathogens may be among the first organisms to be affected and to show the effects of climate change. Their high reproductive rates, the short generation length and the efficiency of propagation mechanisms, in fact, make the pathogens particularly suitable for a rapid adaptation. One of the first and most important impacts of climate change on human society, therefore, might be related to the influence it will exert on the epidemiology of pathogens and, consequently, to the yield losses caused by diseases induced by such agents.

Going to depth

Carbon dioxide: not only negative effects

Most of experimental work concerns the effects of increased atmospheric CO_2 concentration on host-parasite system. The impact of this phenomenon on diseases is indirect as it is mediated by changes in the host plant physiology and morphology. At high concentrations of CO_2 the host resistance can increase; in such conditions, however, there is also an increase in carbohydrate content that can either encourage the development of some so-called "biotrophic" pathogens, such as rust agents, or inhibit others, such as those responsible for downy mildew.

The carbon fertilization effect, on the other hand, involves an increase in biomass and, therefore, changes in plant cover and microclimate. These changes allow some pathogens that instead use crop residues to survive to have a better chance of surviving between growing seasons. In addition, an increase in size and density of plant cover, combined with an increase in environmental humidity, promotes the development of some foliar pathogens, such as rust agents, whiteflies, leaf spots and wilts.

Time machine to study climate change

Special growth chambers, called phytotrons, are used to simulate climate changes. At Agroinnova we have a set of six: they are big enough to allow the growth even of poplar trees. Inside, not only temperature, but also CO_2 levels can be modified. It is possible to have artificial rain, wind, etc.

A visit into them looks like a trip into the future.

The First Biotechnologist in History

There's no need in guessing the name of the first biotechnologist by searching among celebrities. The first real biotechnologist was, in fact, a plant pathogenic bacterium. I'm going to explain you why and tell you the story of this bacterium.

First of all, *Agrobacterium tumefaciens* is a bacterium that causes in its host plants (rose, peach tree, grapevine and many others) the appearance of crown galls, that are neoformations due to abnormal cell proliferation. Such galls are formed mainly at the level of roots and crown of host plants. One of the typical characteristics of this bacterium, which has made him a real biotechnologist, lies in its ability to "transform" the normal cells of the host plants into tumoral cells. Once this transformation has occurred (the process takes 1–2 days), such cells begin to grow and divide abnormally. This abnormal growth occurs independently by the presence of the bacterium. Normal cells' transformation into tumoral cells is due to the transfer of plasmid DNA fragments from the host plant cell. The pathogenic *Agrobacterium*, in fact, contains the Ti plasmid, that is nothing more than a circular DNA, of which a small quantity is transferred into the plant cell, integrating into one of its chromosomes. The presence of this plasmid has sparked the interest of researchers, as it is a formidable vector of genetic material that, as such, has been used for years by biotechnologists, until smaller plasmids have been developed.

Furthermore, this bacterium has also other peculiarities. Against it has been developed what has been for years perhaps the most successful biological control strategy against a plant pathogen (see box page 64). There is in fact another agrobacterium, *Agrobacterium radiobacter*, which represents a formidable antagonist of the pathogen. A very specific strain, coded K84, isolated by the Australian researcher A. Kerr, is able to control *Agrobacterium tumefaciens* by colonizing the host at root level, which is the site of action of the pathogen. The antagonistic bacterium also produces bacteriocines (substances active against bacteria), capable of inhibiting the development of the pathogen.

M. L. Gullino, *Spores*,
https://doi.org/10.1007/978-3-030-69995-6_15

┌─ Going to depth ─┐

What made biological control so successful?

Agrobacterium radiobacter strain K 84 is still, many years after its discovery, the most effective biocontrol agent for crown gall disease management. Why? For at least two reasons. First of all, this bacterial antagonist has no chemical competitor, since antibiotics are not admitted in agriculture in most countries. That's quite an advantage. Moreover, the antagonist is very similar to the pathogen and has the same ecological niche, occupying the same sites that the pathogen would colonize. If the antagonist arrives before the pathogen, which is possible when plant roots are dipped in a water suspension with the biocontrol agent before the transplant, it is surprisingly effective.

Bio Wars

Agricultural products have always been considered, throughout history, a potential target to deprive the enemy of food resources and reserves. It is no mistery that many countries have been producing biological weapons for years and, unfortunately, some still do, though forbidden.

Since the beginning of 1900 several countries, including France, Great Britain, Germany, the United States, Japan, the former Soviet Union, have carried out research on biological weapons, including some infectious agents on plants. This research continued during World War II, taking into account animal and plant pests and herbicide-active substances. Moreover, what we have seen in the cases of the late blight of potato of 1845–49 in Ireland and the Bengal famine of 1943 has well clarified the impact that an epidemic concerning an important crop can have on a country's economy, with enormous social implications.

In the former Soviet Union the biological warfare programme began in 1928 under the Red Army. At the time of its maximum development, more than 30,000 military and civilian personnel worked on it, distributed in 55 institutes and research centres. A large part of their work concerned the production of *Puccinia recondita*, causal agent of brown rust on cereals. The Russian programme also worked on the production of *Puccinia sorghi*, tobacco mosaic virus (TMV), wheat streak mosaic virus (WSMV) and several potato viruses.

At the end of 1930s dates back the US Research Programme, that chose to work on numerous phytopathogenic agents right for their high aggressiveness against economically important crops: in particular the cultivation of rice in China and wheat in Ukraine were considered as possible targets. Between 1951 and 1969 the United States produced and stored more than 30 tonnes of spores of *Puccinia graminis* f. sp. *tritici*, pathogen capable of infecting wheat, causing stem rust, and in 1996 more than a tonn of spores of *Pyricularia oryzae*, causal agent of the so called rice blast disease. These are huge quantities of potential infectious agents, capable of destroying thousands and thousands of acres! North Vietnam's rice fields were also a potential target for Americans in the 1960s. Moreover, there are conflicting reports as to the existence of an American Intelligence programme that in the same

© The Author(s), under exclusive license to Springer Nature Switzerland AG 2021
M. L. Gullino, *Spores*,
https://doi.org/10.1007/978-3-030-69995-6_16

years secretly aimed to wipe out Cuban sugar cane and tobacco crops. In 1960, President Richard Nixon called for the closure of biological warfare programmes, and pathogen spore reserves were officially destroyed.

German programmes on biological weapons, on the other hand, dealt with large-scale production of *Phytophthora infestans*, causal agent of potato blight, *Magnaporthe grisea*, responsible of helminth infection of rice, and of two wheat rust agents, yellow (*Puccinia striiformis*) and stem rust (*P. graminis*).

During World War II, France envisaged the use of biological agents as possible weapons against German potato crops, mainly producing *Leptinotarsa decemlineata* (known as potato beetle) and *Phytophthora infestans* (potato blight).

Similarly, in Japan, while the interest was more on biological agents threatening human health, large amount of cereal rust spores was produced and stored for its possible use on American and Russian wheat fields.

In short, one could just say, for each country its biological war, depending on the enemy to fight. Obviously, the various countries involved in biological warfare not only produced large quantities of biological weapons, but also studied how to convey them: for example, it was proposed to use turkey feathers covered with rust spores to infect cereal crops.

In 1972 the Biological Weapons Convention on the Prohibition of the Development, Production and Stockpiling of Bacteriological and Toxin Weapons and on their Destruction committed signatory countries to suspend their offensive bio-weapons programmes. The former Soviet Union, however, did not stop its own that, on the contrary, grew dramatically in the years 1970–1980 and was officially suspended only in 1992 by President Boris Eltsin.

It must be said that, despite the presence of so many research programmes on biological weapons active against agricultural crops, fortunately it seems that there have never been practical applications. However, at present, research on this topic is very much up-to-date and needed (see box page 67).

An effective European–American cooperation

My interest in biological warfare as a possible threat in agriculture did originate thanks to the attention devoted to this topic by the American Phytopathological Society (APS), the scientific Society that gathers thousands of American and International plant pathologists. Among them, Jacqueline Fletcher and Jim Stack did indeed inspire my work on plant biosecurity in Europe. We have been able to effectively work together in several international projects.

Spores of My Life

This second part of the book deals with plant diseases that I have directly encountered through my work. Spores of my life, therefore, with many references to personal experiences.

These spores are related to different periods—"my many lives" I would say—I spent in the Netherlands, in the United States, throughout Europe, in China, in Saudi Arabia, etc. These are often overlapping experiences—if not I should be at least 150 years old!

This part of the book is also populated by many people—colleagues and/or friends—I met in various circumstances all over the world. All of them, somehow, are connected with the study of a certain pathogen and/or a time of my life.

I hope that reading about these unsuspecting fellows will help to provide an insight into the world of research, whose most interesting aspect is certainly the relationship with colleagues and students from all over the globe.

The diseases considered are the most diverse, as well as the crops concerned: grapevine, tomato, basil, wheat, rocket, carnation, etc.

In addition, I have focused on disease management strategies, since the aim of a plant pathologist is not only to study plant diseases but also—especially I would say —find the right prevention strategies and the most effective treatments in order to maintain plants healthy. This part of the book also takes into account general topics, in some cases controversial (such as the use of pesticides) or connected with radical changes (globalization of trade, biosecurity), certainly not to fuel the debate, but simply to inform or better intrigue the readers with a general overview on the subject.

Everything Started with Grey Mould

My adventure as a plant pathologist began in a vineyard in the surroundings of Asti, in Piedmont, in 1978. I had just started a research scholarship granted by Italian Research Council in what was then the Plant Pathology Institute of the University of Torino. I was immediately involved by my supervisor in field activities.

It was September, harvest season of Moscato, a variety of grapevine grown in my Region, Piedmont, which gives the very famous sweet dessert Moscato wine. I was promptly invited to join the group that was taking on the final surveys of the field trials carried out to evaluate the effectiveness of some fungicides to control *Botrytis cinerea*, causal agent of the infamous "grey mould" of grapevine. This rot greatly reduces the quality of wine grapes and table grapes and, in the case of the former, negatively changes the organoleptic characteristics of wines.

I remember with nostalgia the first field trip. At the wheel of the van of the Institute—a half a bit rickety, very instable because of the height—there was the legendary Dr. Giovanni Gullino, my namesake but not related, whom we always remember for his infinite goodness and kindness.

My first "spores" were the ones of *Botrytis cinerea*, which rose abundantly, just shaking lightly the control bunches that were not treated with the products under test, held as a comparison.

All those spores, which indicated that the trial "had gone well", that the very high disease incidence in the untreated plants would make the field trial successful and the results interesting, aroused my enthusiasm.

In the evening, on the way back to Torino, professor Angelo Garibaldi, my advisor, explained to me, with much kindness and as much firmness, that exulting for the serious attacks of a pathogen in front of farmers is not a good idea, as well as it is not polite to refuse a glass of wine from them. The problem was that the farms hosting our trials visited every day were many, and as many were the glasses of wine kindly offered to us, also as a sign of gratitude for choosing their vineyards for our experimental trials!

I quickly learned the lesson and already the next day, visiting other farmers, I knew how to behave, containing the joy in seeing again so many *Botrytis* spores

M. L. Gullino, *Spores*,
https://doi.org/10.1007/978-3-030-69995-6_17

and starting to appreciate good wine. Fortunately, dr. Giovanni Gullino was driving the van so I could practice with. Within a week, I was very well trained not only in grey mould but also in wine-tasting.

I miss so much my years as a scholarship fellow! I am not ashamed to say that I was happy and completely realized, as I felt appreciated and, above all, encouraged to do my best, even if I was as the low man on the totem pole.

My research topic was precisely *Botrytis cinerea* and of this pathogen, causing very serious damages on many economically important crops (in addition to grapevine, strawberry, tomato, kiwi, rose, cyclamen etc.), I studied aspects of biology, epidemiology and disease management. I always took care of the bad *B. cinerea*, without ever working on the good mould, responsible for noble rot (see box page 73).

Prof. Garibaldi soon involved me in other studies, which ranged over different topics, and I followed him with pleasure, eager to learn new things through the guidance of such a competent and kind person (see box page 74). After 42 years, I keep unchanged the desire to learn, and grapevine remains for me a very special crop, not only for what regards grey mould!

After *Botrytis*, I focused my attention on other important pathogens. Visiting several innovative farms gave me the chance to meet passionate winemakers in different part of the world and collaborators who have remained such over time (see box page 75).

Daughter of agricultural entrepreneurs, graduated in Biology, after almost two years spent in Botany, I found in Plant Pathology "my" discipline, as it offered me the opportunity to do research useful to farmers. In short, it was a branch of research in which I found and still daily find significant social implications and whose application possibilities and practical repercussions seemed close to me.

I have seen many spores, but those of *Botrytis* still manage to arouse a deep emotion in me, bringing me back to my first experiences as a precarious, but happy researcher.

To make Sauternes wines, it takes a special fungus

Botrytis cinerea is a normally feared pathogen because, as we have seen, it is the causal agent of a very serious disease, grey mould of grape, which causes the appearance of rots that alter the quality of affected crops.

Nonetheless, some wait anxiously the arrival of *Botrytis cinerea*, for its role for winemaking, especially of prized varieties such as the well-known Sauternes, in the region of Bordeaux in France as well as in other few areas in Germany and in California. In this case, though, we are not talking about grey mould but of "noble" rot. The particular pedo-climatic conditions of some areas, unfortunately geographically limited, allow the grapes to be attacked in a very light way by the fungus, so that they become only partially rotted and are able to produce particularly fine and sweet wines. In these areas, harvesting is carried out several times, picking the single raisins as soon as they reach the right point of infestation. Now you know how to make Sauternes and why it's such a prized variety.

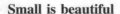

Small is beautiful

Today we work in large departments in which all disciplines are represented. At the time of my first scholarship, however, there were many small Institutions, mostly linked to different Academic disciplines. I remember with great nostalgia the Institute of Plant Pathology, consisting of a few professors. Professor Castellani, who was about 80 years old at that time, a great expert in tropical crop diseases, regularly attended the Institute, after having directed it for years. Both his age and his bearing reminded me of my maternal grandfather: he was like a severe grandpa, though always ready to review our research work very critically and to provide prompt advice and suggestions on the setting of our experiments. And then the Director, Alberto Matta, extremely sober and very authoritative person. Interested in the mechanisms by which pathogens attack and infect plants (physiological plant pathology), professor Matta would try to transmit to all his students his extremely critical approach. Last but not least, there was Dr. Giovanni Gullino, my namesake but not related, fellow of infinite goodness and kindness, with a deep knowledge of the practical aspects of plant pathology. My supervisor was Angelo Garibaldi, the youngest of the group of professors, then. Born on the Liguria coast, he was and still is a great expert on diseases of ornamental crops. I remember how he tested my limits with the first field activities. After all, I was not an agronomist, and a woman working in what was then a purely masculine environment. The first field trips in the countryside, in the vineyards near Asti, were certainly useful to test my physical resistance, my willingness to climb the hills and, why not, my agility. We spent the first weeks always in the fields, to check our trials, up and down the vineyards, and I was always the one with the most impervious side to climb. On the way back, while Giovanni Gullino drove the van prudently, we would start to set the data, which the same evening would be statistically analysed (without the programmes that we use now) in order to have the tables ready for the next day. Fortunately, I passed the test and never, not even for a fraction of a second, did my supervisors make me feel somehow unease. After 45 years, we are still working together. This is how my adventure as a plant pathologist began.

From the Rabinos Farm to the Opus One experience

For over 30 years I had the privilege to carry on experimental trials in the Rabino farm, located in the Roero region, in Piedmont. Wonderful vineyards owned by a very special family: the Rabinos. I got to know three generations of them, from the late grandfather Tommaso to the new generation of the three sisters. I found their tasty wines (Nebbiolo, Barbera, Moscato) all over the world, during my trips.

I'll never forget, in the early 1990s, a wonderful trip to California, with other agronomists and plant pathologists, fully devoted to learn more about its wine industry. Beautiful vineyards, incredible estates, extraordinary red wines, with the unique experience of the visit to the Mondavi's Opus One Winery—like a Sanctuary! —that would later merge with the Mouton Rothschild Group. Drinking Opus One in the San Francisco Bay, on a luxurious boat, during the 4th of July celebrations, with extraordinary firelights. Absolutely unforgettable!

The Netherlands in My Heart

In 1979 I had been working for a few months at the Institute of Plant Pathology and already Angelo Garibaldi and Alberto Matta were proposing me to apply for a scholarship abroad. I could not believe it! Their advice was evidently not to assume I would win a position on the first try, since I was still young and had few academic titles, but it was worth taking a shot.

We carefully selected the foreign Institution where I could carry out a research on resistance to fungicides (see further below): I needed to look around also to work on new research topics, to learn techniques other than the ones carried out at Torino. The one we finally chose was Wageningen, in the Netherlands: an entire University devoted to agriculture. It was a small town, the size of my Saluzzo hometown, which practically coincided with the University campus. I was immensely surprised and glad when I knew that, even though I ranked eighth, I had been selected for one of the eight junior scholarships up for grabs.

I had to change my quiet lifestyle, leave my friends, say goodbye to little Vittorio, son of Laura, my dearest friend, who at three years old considered the Netherlands a very distant land and wouldn't let me go. I overcame the fear of flying and left. I was so thrilled! And so, from 1980 to 1982, I lived my Dutch period.

They were intense and beautiful months. Working with Johan Dekker, the major international expert in fungicide resistance, was a unique experience. He was attentive, open-minded, an authentic gentleman. Moreover, having spent six months studying in Italy, he could well understand our country and its people. I felt immediately welcome and at ease.

I learned so much in that period! My research focused on the resistance to a new family of fungicides called ergosterol biosynthesis inhibitors. I learned a lot not only from a technical point of view: I helped in fact the Dutch colleagues in practical teaching and gave my contribution in carrying out exercise-lessons, field experimental activities. I had the opportunity to support my research with a lot of laboratory practice. Appreciating my enthusiasm and curiosity, the Dutch colleagues involved me in many activities (see box page 79).

It was an exciting time. I experienced the importance of what they had taught me in Torino: networking with the best laboratories, taking great care of relationships with foreign colleagues. Even today I can count on very fruitful collaborations with the Dutch lab where I worked in the early 1980s. People have changed, but friendship and good feeling are the same. And Professor Dekker, with whom I used to have interesting discussions, has been for many years a point of reference throughout my scientific career.

Moreover, there was the personal side of the experience. I lived at the International Agricultural Center (IAC), a very well organized facility designed to host foreign researchers from all over the world. We immediately formed an international group of about fifteen close friends.

In the long and clear summer evenings, after dinner (at 6 p.m.) and late study or work sessions, we would go cycling until midnight, cross the Rhine with a boat that had a special charm for us and ride happily in the town of Rhenen, on the other side of the river, in front of Wageningen. We would share dreams and ideas.

After 40 years, I am still in contact with some of those friends. We founded the so called 5th Floor Association, named after the college where we lived (the International Agricultural Centre). I had research projects with Wohlert, the German friend who bought my Dutch bicycle when I had to come back to Italy; I often keep in touch with Rui, the Portuguese friend working in the environmental field, and with Quique and David, entomologists in Argentina and the United States, respectively. And then there is Clara, researcher from Bologna who chose to stay in the Netherlands, to whom I am bound by a great friendship. We have followed different paths and made different careers, but we all have found our way in our specific areas. And we are planning to meet in 2021 again! (See box page 80.)

The Dutch experience has left its mark on each one of us. Living in such an international context was extremely enriching for a person like me, who came from a small town, and gave me the push I needed to follow an international path.

The two white Samsonite suitcases I received as a gift when I left for the Netherlands became the symbol of my nomadic life (see box page 81).

Every time I return to the Netherlands I feel nostalgic for those carefree years and grateful for a system that certainly helped me to grow.

A lab full of music

At Wageningen, I did work in the lab of Maarteen De Ward. Close by, there was the lab of Adrien Fuchs, a very much extroverted researcher. In his lab there was Fritz, a very talented technician who used to sing opera all-day-long while working. I still remember that lovely voice accompanying my experiments. A very unusual atmosphere in the very sober and cold Dutch system!

The 5th Floor Association

Beside the working experience, the Wageningen period was also a wonderful life experience! We were a group of young researchers, from all over the world. Working in different labs, on different topics. Most of us lived at the International Agricultural Center (IAC), in the mid of this small University town. Also called, jokingly: International Agricultural Cemetery. Most of us lived at the 5th floor. The doors of our rooms were always open. After full days in the lab, inedible dinners at the restaurant, early evening again in the lab or studying, at 9 p.m. we would gather in the Bar, also called Library at the upper floor. Back to our countries, we all made our own ways: David Onstad, entomologist in the US, Rui Cortez, environmental chemist in Portugal, Antonina Wasilewska, a physicists in Poland, Marcelino Freyre (Quique), entomologist in Argentina, Zofia Szeitchtenberg, horticulturist in Poland, Angelica Bianu, horticulturist in Rumenia, Alexandridis Christos and Michael Papadimitriu working in Greece, Kathy Lefevere, from Belgium, now living in Australia, Katharina Scholz, in Germany. Only Clara Marcucci, horticulturist from Bologna, left research, becaming a real Dutch. After almost 40 years, we are still in touch and we are planning to meet as soon as the Covid pandemics will be over.

A collection of suitcases

For more than 30 years my home was a suitcase. Or better, a lot of suitcases, because I have received many of them as a gift over time and I have literally consumed them.

Receiving a suitcase as a gift has always aroused deep emotions in me. In fact, for those who travel a lot the suitcase is a precious and much loved object to take along. Luggage is useful to put inside our beloved things and later the new items that we will probably bring back as good memories of the journey; moreover, it reminds us of the giver and of the fact that our stay away soon or later will end.

I have never thrown away a single suitcase, even when they were worn out by too many trips, because each of them represents for me a piece of life. The two white Samsonite that I got as a present for my first stay in The Netherlands are still there, as beautiful as the first day, despite being used a lot. The secret to keep your Samsonite beautiful? Massage them with Nivea cream after washing, to keep them elastic and shock-resistant.

Over the time I've learned to reduce the contents of my suitcases a lot. And I started using smaller ones, though still able to contain not only basic items, but also everything needed to best represent Italian style and elegance throughout the world. Because, after all, I'm Italian!

America, What a Passion!

After the Netherlands, it was the turn of the United States of America. It was 1982, and as soon as I came back from the Netherlands, I started to apply for another scholarships, looking forward to my next experience abroad. This time it would be out of Europe.

A National Research Council notice had just come out that seemed to be made especially for me. I chose my next destination of study looking overseas. After all, I have been dreaming of America since I was a child, yet suddenly it seemed to me so far away… Anyway, there was no way to convince my supervisors that maybe England would be a good compromise. Today I can only be very grateful for their firmness about destination.

This time I ranked halfway in the final merit list for scholarships abroad and in November 1983 I began my American adventure.

I left Europe in tears, but managed to adapt very quickly to the new context, so much so that I cried even more at the end of my overseas experience, nine months later, at the time of returning to Italy. I cannot forget the arrival at Dulles International Airport, in Washington D.C., so lost in the countryside around the city that for a few minutes I thought I got to the wrong destination. This was certainly one of the many occasions in my life that made me feel a clumsy girl from the suburbs: simply the airport of Dulles was already equipped with a pneumatic bus for the transport of passengers from the terminal of arrivals to the main one.

When at last, after having crossed what seemed to me an endless plain, I reached the terminal and saw my Uncle Piero waiting for me, I felt at home. I remember Uncle Piero's stern comment when I told him "How nice, I feel like I'm at Saluzzo!"

My American adventure began, muffled during the first week by the stay at my uncles' home, by aunt's roasts and good bottles of Barolo and Barbaresco wines by Pio Cesare. Uncle Piero was about to finish his American career and in 1985 he would return to Italy: I had the honour to help him to empty the cellar of the last stocks of wine, all from Piedmont (see box page 85).

Of my first impact with Washington I remember the impression I got from the big roads and boulevards without parked cars, before realizing that in the United States you don't park like in Italy. And then the high percentage of obese: while I was queuing for the registration of the second hand Fiat 128 just purchased, I was struck by the number of persons of size around us!

Within a few days, with the help of my uncles and my American professor, Hugh Sisler, I settled completely: car, house, phone.

I spent the first week in Bethesda at the uncles' home, where Prof. Sisler would take me back in the evening and stop for a chat with Uncle Piero. They talked about everything: Hugh Sisler was tall and lanky ("kind of cow-boy, like John Wayne", remarked Mr. Rabino, when I went with Prof. Sisler to visit his company) and had a lively intelligence. His thoughts ran so fast that words could not easily keep up. He spoke quickly, with his mouth closed, and it was very difficult to catch the point and follow his reasoning. Sitting in an armchair, he would talk and gesture as he held a cup full of coffee on his knee. Yet he never poured a drop of coffee on the magnificent carpets of my uncles.

The first days in his laboratory were very hard: I had the impression of not knowing English sufficiently, as after 5 min of discussion with Hugh Sisler I was already lost. I felt better when I met his wife and had it clear that in fact nobody in his family was able to follow his thoughts, nor even his words. I was advised to let him speak. Since then I did my best to follow his explanations, but I no longer felt so unconfident when I didn't understand something well.

Hugh Sisler was a world expert on fungicides: I worked with him on the mechanism of action of dicarboxamide fungicides, which were used at that time against grey mould.

I went to live in College Park, sharing an apartment with Shirley, a typical American girl who helped me to deepen my knowledge of the country.

Working in the lab, I learned many new techniques, dealing with a very innovative topic. Life on campus was less fun than life in Wageningen, also because wandering around after a certain hour at night was unsafe.

Maryland University could only be reached by car, driving along a busy Beltway, where my FIAT 128 (see box page 86) often was jolted when it was caught between two tracks. Sometimes I felt so unsafe driving such a small car that I missed the exit of the ring road and chose the next one, terrified by the huge trucks overtaking me from both sides. On Beltway I could well realize the boundless spaces of United States: I missed so much my bike rides in Holland! However, I immediately dived into American life. During the day I worked hard in the lab, where interesting results arrived soon and where I made many friendships. Washington D.C. is a beautiful city, culturally dynamic and, interestingly for a young scholarship student, inexpensive.

The closeness to the uncles, which I visited almost weekly, meant that there was never any homesickness. America fascinated and conquered me.

I returned to Italy just in time to organize the new American scholarship, this time at Cornell University.

After many years and two other long study experiences—at Cornell University and at Pennsylvania State University—I have sometimes critical feelings towards United States, similar to those described by the late writer and journalist Oriana Fallaci, recognizing in this country and its inhabitants many positive aspects and some, instead, more questionable. But I have always considered the positive ones largely prevalent.

The United States is the perfect country for those who have a great desire to work, especially for those like me who appreciate healthy competition. It is a country where anyone can emerge, if they have a good idea. It is in this country that I have always found inspiration for the most innovative research topics.

Living and working in the United States in different periods for a total of four years contributed to shape the one I am now, for better or for worse. In America I have learned not to underestimate myself, and understood that you can always start over. I have learned to dare and believe in myself.

I am very attached to United States, thanks to very solid friendships that I tightened (see box page 86) over the years, the great respect I have for this country and its Institutions.

In 1983–1989 I really lived my American period: after my wonderful years in the United States, I wouldn't miss the chance to fly overseas.

Now trips to USA are less frequent, but I am still intensely bound to this great country.

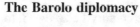

The Barolo diplomacy

Italian people are known worldwide for their sense of hospitality. I must tell that I learned from my mother the art of hosting friends at my house. It is always pleasant to get a chance to better know our colleagues from around the world, sharing a pasta dish, a good brasato, all associated with a glass of good Italian wine. I'm missing so much such opportunity during this period…

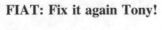

FIAT: Fix it again Tony!

Driving my Fiat 128 in Washington I realized the American's very low regard for FIAT cars in 1980s. The Italian brand was considered, in fact, the acronym of "Fix it again, Tony!", meaning that the cars of such brand continually needed maintenance. Actually my 128 had a lot of problems. My "Tony" was Professor Hugh Sisler who, thanks to a manual skill equal at least to his scientific talents, was able to repair the car every time I was in trouble. Finally today, in the post-Marchionne era, Americans have changed their mind about the Turin brand that boasts a tradition of 130 years.

Friends for life

If the United State is my elective country I also owe it to Charlie Delp, dear and precious friend and researcher in the private sector, "father" of the benomyl fungicide. I met Charlie in the Netherlands in 1980 and since then an intense friendship has begun, reinforced by the common interest in the study of fungicide resistance. Charlie has been for years a patient reviewer of my papers: his punctual and severe editing has certainly contributed to make my English more fluent. Many times he said to me: "Your English is correct. However, we simply do not write like that". How patiently he revised my papers! It was Charlie to give me interest for the International Society of Plant Pathology (ISPP) and to help me—now I can reveal it—to prepare my successful bid to bring the 9th international congress of the Society, ISPP 2008, in Torino. At the age of 94, Charlie is still very active and brilliant, and participates intensively and critically in the activities of ISPP.

All at the Seaside!

Thinking about spores of my life I cannot forget the hundreds of experiments carried out in Liguria, on the Riviera Coast, at two research centres located respectively in Sanremo and Albenga. The former was a research centre on floricultural diseases, while the latter was an experimental centre of the Chamber of Commerce of Savona.

The old Institute of Plant Pathology of Via Pietro Giuria in Torino was lacking space to carry on experimental research, as the kind of tiny greenhouse built in the courtyard certainly did not allow much practical trials. In Sanremo and Albenga, on the contrary, we found suitable facilities, such as modern greenhouses and experimental fields, which allowed a perfect 40-years long collaboration between the plant pathology group from Torino and Ligurian farmers.

It was a few months after the beginning of my career as a scholarship fellow, and I could be really proud of being already involved in experimental studies in Liguria, where I started my trials on anemone. I would go there at least once a week, driving my old A 112, loaded with all sorts of materials: various pathogens grown on artificial substrates used for artificial inoculation of plants, plastic bags, plant protection products and several microorganisms. We would leave at dawn, driving along the Torino-Savona highway, on whose Apennine section overtaking was mainly forbidden. At the Albenga experimental Centre we would work at very intensive pace, with the technician Francesco Azili who ran all the activities with care and passion, keeping greenhouses and experimental plots as they were his personal garden. In Sanremo, we worked with Carlo Pasini and Ferdinando D'Aquila (see box page 88).

The activities carried out in Liguria, intended as support and integration to our research in Torino, proved to be useful for local farmers and pioneered the agricultural development of the region.

The beginning of my experimental trials in Liguria, in the early 1980s, coincided with the development in the plain of Albenga of floriculture, which gradually replaced horticulture. In the late 1970s-early 1980s I had the chance to witness the golden age of Ligurian floriculture: carnation, rose, anemone were at that time the

M. L. Gullino, *Spores*,
https://doi.org/10.1007/978-3-030-69995-6_20

main flower crops. Excellent breeders started to produce (and still do) the most beautiful and appreciated cultivars of carnation grown worldwide (see box page 89). Before marketing them, their susceptibility towards *Fusarium oxysporum* f. sp. *dianthi,* a very dangerous pathogen of carnation, worldwide, must be tested.

In the 1980s flowering plants in pots (daisy, cyclamen, etc.) began to replace carnation, rose and cut flowers, leading to hundreds of experimental trials aimed at identifying and controlling pathogens' attacks on those crops. In 1990s and 2000s research focused on new flowering species and varieties (especially potted plants), with extremely high diversification. This is something that can really complicate a grower's life, but at the same time is compelling and fascinating for a plant pathologist. It can be said that, in floriculture, a new disease per day gives researchers and technicians the right thrill to carry on their work (see box page 89).

Nowadays, while the cultivation of the most varied species of potted flowers continues, also medicinal plants have become very popular, so that, as plant pathologists, we keep on studying new pathogens affecting this special kind of plants. Very often the work on ornamental plants involves American colleagues (see box page 90).

I forgot to specify, but it was obvious, that in these trips to Liguria, the sea was only seen by the car window, from the highway...

Very special technicians

Due to its complexity, the floricultural sector requires the widespread presence of well trained technicians on the territory. Working in Liguria I met exceptional technicians. Bernardino Armato, with his perfect knowledge of the crops and the territory, made Cooperativa Ortofrutticola of Albenga a sure point of reference in the 1970s. The same can be said of Giorgio Bozzano, my own age, taciturn and competent fellow, who, after Armato, kept on the tradition of Ortofrutticola. There are no words to describe Francesco Azili, who for almost 40 years has managed the Experimental center of Albenga, working with great competence and infinite dedication. I remember the experimental plots defined by Francesco with incredible precision and the yellow cards that indicated the different treatments, perfectly aligned. And the kind severity with which Francesco Azili, shaking his head with gentleness, reproached disorganized and careless students.

Curiosity

Ligurian breeders, the best ones in the world

Liguria is not only a land of sailors, but also of very good breeders, that is of expert in obtaining, through wise crossings, varieties of flowering plants, much appreciated for their aesthetic and agronomic characteristics. Being a breeder is mainly a family business and Liguria has been the birthplace of families (Mansuino, Baratta, Brea, Sapia, Nobbio, Santamaria…) of important breeders who have "invented", I think we can say, the most beautiful varieties of rose and carnation in the world. Everywhere carnations and roses are grown worldwide, you will find the wonderful cultivars developed by the Ligurian breeders.

Going to depth

Floriculture, fertile ground for plant pathologists

Floriculture is a very interesting economic sector, characterized by a series of peculiarities that make it both challenging and fascinating. It is in fact an extremely dynamic sector: crops and their varieties change continuously, according to consumer demand. Flowering crops are often intensively grown in more or less advanced greenhouses, where they rapidly succeed one after the other. A very favourable system to disease appearance.

The American colleague working on flowers

During my stay at Geneva, Cornell University, I had the pleasure to meet Margery Daughtrey, a very well known expert on diseases of ornamental crops. Since then, it started a very intensive cooperation and exchange of information. Very often we work on the very same diseases in two different parts of the world. The exchange of experience is thus very helpful!

Coast to Coast (Sanremo–Venice)

Do not think that with this term I refer to the crossing of the United States, from the East to the West Coast, although carried out many times during my American period. What has left its mark on me is, instead, a much more domestic coast to coast.

I told you about Liguria and the thousand trips on the Torino-Savona highway. While keeping on driving my car—which in the meantime I had replaced with a FIAT Uno (the car for me has always been a simple means of transport and not a status symbol)—along the same but upgraded highway, in 2003 thanks to the experience gained in China (see further below) I had the chance of coordinating international courses on sustainable development for Chinese experts.

Where? In Venice. I caught this beautiful opportunity and started to go there frequently. The island of San Servolo, in the Venetian Lagoon, welcomed me in all its splendour—it was a former asylum for the insane, as was Grugliasco, the seat of Agroinnova (a coincidence?). Overcome the anxiety from high water (see box page 92), that often occurs in Venice making complicated every kind of transfer, I learned to appreciate this wonderful city not only for its charm but also for its inhabitants. Kind people, able to live calmly, appreciating to the maximum the beauty and the character of their city, as much enchanting as demanding.

Thousands of students from China and Eastern Europe, have come "to learn" what sustainable development could mean, declined in different ways and areas. Topics included economics, agriculture, natural resources, architecture, transportation, all seen from the perspective of sustainability. The courses included a practical part, involving Italian Universities and companies, in order to foster collaboration at all levels. The courses were mostly held at the Venice International University, a Consortium of Universities from different countries, started with great foresight in the 1990s, and partially in Torino, at Agroinnova. The experience gained in planning these courses, together with the years of work in many developing countries with several research and technology transfer projects (see page 96), have enriched my knowledge of these countries—especially China, caught right in its phase of fast development—and made me appreciate their many positive aspects, though recognizing the weak points.

Going to Venice has always been a great pleasure—even though there were no high-speed trains—for the emotion of entering the station and seeing so much beauty unfold before my eyes; for the professional satisfaction; for the good quality of life in such city.

Now I no longer fear high water and, when I am in Venice, I not only accept but also appreciate the placid pace that water gives to this city.

High water: will it end thanks to the Mose?

Venice has been since long time characterized by the presence of the so called "high water", caused by exceptional tide peaks that occur periodically in the northern Adriatic Sea. The peaks reach their maximum in the Venetian Lagoon where they cause partial flooding of the city. The phenomenon occurs mainly between autumn and spring, The flooding caused by the *acqua alta* is not uniform throughout the city of Venice because of several factors, such as the varying altitude of each zone above sea level, its distance from a channel, the relative heights of the sidewalks or pavements (*fondamenta*), the presence of full parapets (which act as dams) along the proximate channel, and the layout of the sewer and water drainage network (which acts as a channel for the flooding, as it is directly connected with the lagoon). To assist pedestrian circulation during floods, the city installs a network of gangways (wide wooden planks on iron supports) on the main urban paths. This gangway system is generally set at 120 cm above the conventional sea level, and can flood as well when higher tides occur. If you are not in a hurry, the "high water" is quite an experience! I got organized and kept a pair of boots in my usual hotel. Just in case! The Mose, is an integrated system consisting of rows of mobile gates installed at different points, able to isolate the Venetian Lagoon temporarily from the Adriatic Sea during "high water" tides series.

Torino-Beijing Round Trip

When at the end of 1999 I started working with some Chinese Institutions I found, to my great surprise, a country completely changed compared to the one known only five years before, during a pleasure journey. Brand-new airport, no more bicycles everywhere, skyscrapers popping up like mushrooms. At the beginning of 2000 began an intense collaboration with Chinese Universities and Academic Institutions aimed at research, training and technology transfer in the field of sustainable management of plant diseases. Agriculture has always been one of the productive areas on which Chinese government has most invested, due to the long history of farming and the huge rural population of this country. The importance of agriculture in the economy is measured as the value added of the agricultural sector as percent of GDP. In 1999 agriculture accounted for 16% of Chinese GDP and 67% of the workforce were employed in this sector, while in 2019 agriculture accounted for 7% of Chinese GDP and 25% of the workforce were employed in this sector. Just for a comparison, the average percent of agricultural contribution to Italian GDP has declined in the last twenty years from 2.72% to 1.93%, while the percentages in USA were respectively 1.6% in 1999 and around 1% in 2019.

The projects carried out (see box page 95), in the frame of a very intensive cooperation between the Italian and Chinese governments, aimed at promoting the Italian model for sustainable agriculture, plant and environment protection, healthy and safe food, with the consequent promotion of a regulatory framework, and strategies and technologies able to ensure environmentally friendly production. In agriculture consumer health and environmental protection are strictly connected.

Like environmental protection, consumer health protection is also an issue that needs to be tackled globally, requiring targeted interventions across borders. Internal pressure for more and more stringent controls on goods imported from China and for compliance with health and quality standards must be supported in China by targeted measures. Attention was also paid to improving the quality standards of agro-food products, especially in response to increased domestic demand for organic products and as a means of encouraging exports (see box page 96).

The entry of China into the World Trade Organization (WTO) a few years ago provided great commercial opportunities for national agriculture, actually still difficult to seize because of a rather outdated production system that fails to meet the quality and health standards of foreign markets, especially of European ones. At a legislative level, the response was to develop regulation to promote regional product characterization, the introduction of quality labels, and the establishment and strengthening of health and quality control structures. The product distribution chain has also been strengthened and improved, starting from the processing capacities of the agro-food industry.

In the field of training, the aim has been to disseminate up-to-date technical and scientific knowledge, to train staff capable of internalising innovations and leading change in a sustainable perspective, both as experts in the industrial sector and as public officials or academics.

In the process of agricultural modernization the human factor has not been neglected, being aware that the presence of trained and qualified professionals is the true engine of every transformation (see box page 97).

In this context cooperation with Europe and especially with Italy is particularly appreciated: China recognizes the great ability and the innovative potential of Europe, from the regulatory, technological and social point of view. My hope is that from the Old continent may come not only threats of custom measures against the "Chinese competitor", but also the voices of those who are aware that China is a great market opportunity for strategic investment in research, innovation and development.

The project activities have been an opportunity to set up a very solid and long-lasting partnership with Chinese research centers and academics (see box page 97). The previous activity of cooperation and technology transfer, initially aimed at solving practical issues related to plant protection, very soon verted towards research projects that saw an intense collaboration between European universities and research centers with the best Chinese academics (see box page 98). Food safety— such an urgent issue in China!, biosecurity and reduction in the use of obsolete pesticides are some of the topics covered over the years. An intense exchange of students between Italy and China has allowed build up good collaborations.

Going to depth

Crucial matters

The cooperation programme, started in 2000, covered a country (China) and issues (sustainable agriculture and food security) that are extremely crucial in the international economic and political agenda. The co-financing of projects to a substantial extent by European and Chinese partners, their inclusion in the framework of wider international cooperation projects, have ensured their efficiency and effectiveness.

The overall objective of the programme was creating a partnership between Universities, Chinese and European Agencies and companies able to cooperate in the field of sustainable agriculture to ensure the quality and safety of Chinese agricultural production and the protection of the environment. Italy and Europe have much to teach and share with China in the agricultural sector, in terms of research, training, technological innovation, quality production and environmental protection.

The activities carried out concerned training, technology transfer and research, especially on fresh-cut vegetables or minimally processed products destined to Italian tables. The ultimate objective was not, in fact, only to reduce the impact of Chinese agriculture on the global environment, but also to protect the health of Italian and European consumers from a distance, trying to align Chinese products destined for foreign market with European models and especially with the Italian one.

The collaboration with China has obviously taken into account the evolution of the Chinese economy: today the Chinese work with us according to current partnership models in the European Union or directly supporting projects.

Going to depth

Technology transfer and research

In the technological sector, field trials carried out in Inner Mongolia and Shandong provinces on tomato, in the Shanghai region on pear, in Hebei province on strawberry and in Xinjiang on grapevine, have proved the technical and economic feasibility of drip irrigation systems, the convenience of automated fertilizer and agrochemical dosage, the effectiveness of biodegradable mulching films to reduce soil contamination by plastic, the importance of adopting integrated pest control strategies to reduce the use of pesticides and the consequent presence of their residues in food.

Tolerant and resistant varieties, grafting on resistant rootstock and biocontrol agents are among the control measures identified as able to permit the reduction of chemicals. The use of bio-fertilisers and soil improvers has also been encouraged for preventing plant diseases. Organic farming, fully consolidated in 2000 in Italy and Europe, was not yet, in the early 2000, a priority in the framework of Chinese agricultural development: collaboration with Europe has certainly been useful for the Chinese. A specific project funded by the European Union permitted to start Agricultural Courses in Organic Farming in five Chinese Universities.

Curiosity

Training

Wide-ranging interventions have been carried out for the dissemination of up-to-date technical and scientific knowledge, and for the training of personnel capable of internalizing innovations and guiding change towards sustainability, both as industrial experts and as leaders of the relevant administrations. The exchange of experience and technology with industrialized countries is considered by China as a key factor to speed up a restructuring process that, if faced only at a national level, would be extremely long.

Curiosity

Some figures

Between the end of 1999 and 2014 I travelled more than 100 times in China, for mostly very short trips to Beijing or Shanghai, going on Saturdays and returning on Wednesdays, interspersed by longer stays in the Chinese countryside for field trials. A dozen Chinese students chose Agroinnova for their Ph.D programme, while dozens of technicians and students came for shorter visits.

Beijing hosted the International Congress of Plant Pathology in August 2013, welcoming 2,000 participants from all over the world.

Nevio 'the Chinese'

China, October 2001. I am in the Xinjiang region –
kind of a small Piedmont - to organize the first
collaboration projects. At the end of the day, they
take me out for a tour of wineries. At the other end of
a cellar, while I'm tasting all but exceptional wines, I
notice someone who can only be Italian. In fact, it is
Nevio Capodagli, much more than Italian! I want to
get in touch with him. I didn't know that this chance
encounter would lead to an endless collaboration.

Nevio had graduated in Venice and left for China
in the 1980s to improve his Chinese—perfect, by all
accounts. During his journey he met Sonia, a
beautiful Chinese woman, from a very good family,
and decided to set up over there, gradually acquiring
a deep knowledge of Chinese culture.

Nevio is since that day in the Xinjiang a great
collaborator of Agroinnova, perfect liaison with
Chinese partners, as well as our reference point in
China. He is now more Chinese than Chinese people
are! Nowadays Nevio changed his field of expertise,
working with football teams.

Inshallah

In my travels around the world, hunting for interesting pathogens, the Arab world could not miss, visited for short periods since the early 1980s and then better known thanks to the numerous projects carried on in Palestine and Jordan in the mid-1980s, in Morocco since 1996 and later in Egypt until more recently in Saudi Arabia.

Also in the Arab countries I have dealt mainly with diseases of horticultural crops, trying to transfer the results of researches carried out in Italy and adaptable in that area.

Very often I have worked in Arab countries as part of international projects, supported by the European Union or international organizations, along with French, Spanish, Greek and American colleagues. In some cases, I ran technology transfer projects on behalf of international agencies (see box page 101).

Every experience in the Arab world has been interesting, since the first one, certainly challenging, in Palestine, where, through a Peace Campus Project, we tried to get Palestina, Jordan, Egypt and Israel researchers, to work together. Not easy: the common denominator was management of fungal pathogens with non-chemical control means. A very innovative project at that time!

Unfortunately, then as now, the goodwill and sincere intentions of individuals were and are often frustrated by very complex general situations that make peace seem an unattainable goal.

The experience in Morocco, which started, as I said, in 1996 and lasted to this day, was much more satisfying. Even in that case I worked on seacoast horticultural crops, from Casablanca to Agadir and beyond, an interesting area where vegetable crops are mixed with citrus plantations. The first projects concerned the transfer of technologies developed in Italy to replace the use of methyl bromide for soil disinfestation (see page 175).

The Moroccan horticulture was at that time going through a great development phase and the local farmers, like all their colleagues in every part of the world, wanted to use the most effective pest control methods. It was not easy to convince the farmers, their technicians and the politicians of the harmful effects of the large use of this pesticide, introduced in Morocco since a few years (at that time) and undeniably effective for the management of soil-borne plant pathogens. Why in the world we should reduce the use of methyl bromide—they argued—a product that you in the industrialized countries have used successfully for many years? It was not easy to answer questions like this and convince everyone that, within a little time and with the help of other countries—Italy in the lead—they would finally succeed.

They were years of intense work: thanks to the commitment of good colleagues and passionate technicians (see box page 102), we managed to help Morocco comply with the schedules internationally imposed to reduce this dangerous ozone-depleting fumigant (see page 175).

Today I keep on working with Moroccan colleagues, also in the area of Agadir: in addition to fungal pest control, in recent years we have developed an intense collaboration on composting strategies, helping Morocco to convert urban, agricultural and industrial waste into a resource.

Once more, the collaboration began with the support of UNIDO (United Nations Industrial Development Organisation) (see box page 103), and then continued with funds made available by Moroccan producers.

Very interesting has been, over the years, the collaboration with Egypt, started in the mid-1980s with the already mentioned Peace Campus Project and continued in different fields, involving both former professors, well established at their Ministry of Agriculture, and younger colleagues at the beginning of their career.

Also very interesting was the project that, although carried out in a quite turbulent time, led seven Egyptian universities, located in poor Nile areas, to set up a specialized degree in Sustainable crop protection, in collaboration with Italy, Spain and Greece. Working with the most disadvantaged areas of Egypt has been a very compelling and formative experience, which has highlighted the great desire for cooperation of this country, as hospitable and welcoming as Morocco, but certainly less projected towards the future and, at this particular time, caught by a series of domestic issues inhibiting the interaction.

More recently, but with similar interest, I approached another part of the Arab world, starting to work in Saudi Arabia, modern and ancient country at the same time, full of contradictions, certainly not very open to the female world, but in transition. A whole new world to explore, along with its crops.

Why working in developing countries?

The experience in Morocco, the first for me in a developing country, has been very useful. As already said, I was born in a small town and grew up in a family of farmers. When I was a child, but even later, it was quite common for me to hear comments, made in absolute good faith by local farmers who criticized any form of collaboration with other countries—such as Spain, at the beginning, and then Tunisia and Morocco —seeing them as potential competitors. Having been raised in a very open environment, I firmly believe in a world that can only be open to any form of exchange. In a globalized world every country has to specialize in doing and producing what it does best: in agriculture there will therefore be areas intended to provide products (fruit, vegetables, flowers), technologies (greenhouses, machinery, seeds etc.) or processes (innovative marketing systems). From this point of view, therefore, it seems to me not only right but proper for a researcher to be open and collaborative to everyone and, in particular, to those countries which, for different reasons are less technological advanced. The benefits will be seen in the long run. Improving its economy, Morocco will be able to offer opportunities to its inhabitants. Europeans, on the other hand, will be consuming fruit and vegetables produced in Morocco according to European standards. And the same kind of collaboration can be extended to other countries.

Mohammed

I met so many Mohammed in Morocco! And so many researchers and technicians are good and passionate in their work. Among them Mohammed Ammati, nematologist (that is an expert of nematodes) at the Faculty of Agriculture in Rabat, with a solid education at the University of Davis (California, United States).

Thanks to a project that we carried out together funded by UNIDO, the United Nations organization that deals with industrial development, Mohammed managed to successfully make use of every single dollar he was given and was able to build up a team of young people working all over the country. This allowed him to get out of the smelly basement full of rats in which the older university colleagues had relegated him, trying to discourage his initiative and resourcefulness.

With Mohammed I travelled all over Morocco and had the chance to know and appreciate the kindness and the hospitality of its people and their habits, as well as the simple and very tasty country food.

Riccardo

Riccardo Savigliano, a former student of the University of Torino, has worked for several years in Vienna, at UNIDO, after having completed a Ph. D and worked for some years at Agroinnova on international projects especially in the field of the replacement of methyl bromide. Competent, calm, polite, Riccardo has acquired a remarkable ability to manage international projects. Today we keep on our collaboration in different parts of the world. He knows Agroinnova is an effective partner and we know he is always an attentive interlocutor.

Torino, the World Capital of Plant Pathology for One Week and Even More

I cannot fail to tell you the story of the 9th International Congress of Plant Pathology, held in Torino, at the Lingotto Congress Center, in August 2008. An unforgettable adventure, started in 2002, with the opportunity, seized on the fly, to bid to bring in Torino the World congress devoted to my research field.

Torino had then just been selected as the host city for the 2006 Winter Olympics. Nothing seemed impossible to us, at the point that, with the help of the Torino Institutions—which collaborated in the preparation of the bid, to provide colleagues from other countries called to express a preference, an image of our city certainly unknown to most foreigners—and of two Scientific Societies dealing with Plant Pathology in Italy, we won even beating the United States of America.

I still remember the phone call at 8.30 in the morning of December 21st, 2002, from Peter Scott, who was at that time past-President of the International Society of Plant Pathology, followed by my shout of exultation that brought everyone to my office. That immense joy made me quickly overcome the embarrassment of beating my beloved United States. The organization of ICCP 2008 began immediately, before Christmas 2002 and involved many people. We all worked with passion and enthusiasm.

The Congress, held at the Lingotto Congress Centre, was a great success, gathering over 2,000 researchers from all over the world: Italy (194 researchers), USA (159), United Kingdom (101), Australia (93), Japan (77), Spain (68), China (67), France (61), Germany (56), the Netherlands (43), Canada (39), Belgium (38), India (38), New Zealand (32) and South Africa (32).

In addition to the scientific interest of the event, that brought together researchers of international renown in interesting debates and discussions, ICPP 2008 amazed its participants with social events characterized by typically Italian elegance and style, such as the Welcome Reception held on the opening Sunday in the evocative setting of the Rectorate Courtyard of the University of Torino, the mid-week dinner in the hilly area of Langhe, and the acclaimed closing concert by a trio of classical musicians (classical guitar, flute, piano) and two opera singers (tenor and soprano) (see box page 106). Wonderful memories (see box page 107).

© The Author(s), under exclusive license to Springer Nature Switzerland AG 2021   105
M. L. Gullino, *Spores*,
https://doi.org/10.1007/978-3-030-69995-6_24

The true engine of the Congress—and guarantee of its successful organization (see box page 108)—was our passion, an essential ingredient of any work, scientific and not.

For seven days in 2008 we relived the emotion of the Winter Games, with the leitmotif "We did it with passion". I have to say that it makes me very happy when, traveling around the world, I happen to meet colleagues who still remember with pleasure the scientific soundness and festive atmosphere of ICPP 2008.

My desire to organize events and Congresses is certainly not over, also because the security of Grandma Severina's "remote" help (see box page 109) gives me an extra boost. The recent Covid-19 pandemic, is forcing us to reshape the way Congresses are organized (see box page 110).

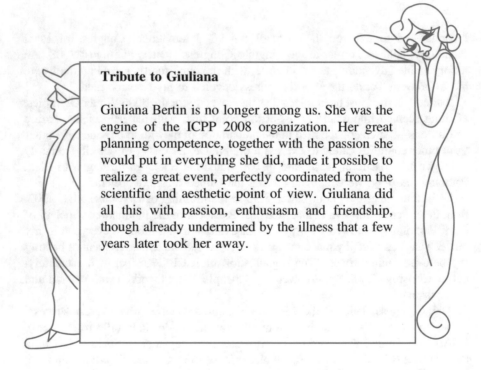

Tribute to Giuliana

Giuliana Bertin is no longer among us. She was the engine of the ICPP 2008 organization. Her great planning competence, together with the passion she would put in everything she did, made it possible to realize a great event, perfectly coordinated from the scientific and aesthetic point of view. Giuliana did all this with passion, enthusiasm and friendship, though already undermined by the illness that a few years later took her away.

Champagne bubbles

Many people asked me how the organization of a Congress like ICPP made me feel afterwards. I took me a while to find the answer: immediately I felt a huge fatigue, resolved with ten hours of sleep. And then a pleasant euphoria, comparable to the sensation that follows a good (but not excessive) stiff belt of excellent Champagne.

After years of German-prussian style organization—something I am good at, as a consequence of the strict education I received—the week of ICPP 2008 turned out to be like a real party, with the presence of many friends and colleagues, and a perfect organization that allowed to live the Congress in a relaxed and safe atmosphere. We were an efficient team, able to cope with any emergency—which fortunately was not necessary. So many pleasant consequences—scientific and not—followed this Congress, together with the satisfaction of having brought a world event right in our country, in my own city and in the Institution where I work.

The Devil wears Prada

Do you remember poor Andy, assistant to the terrible Miranda in the move "The Devil wears Prada"? Well, during ICPP 2008 it was up to Carlotta Bianco being Andy, assisting me continuously, so as to solve all the problems and allow me to live in complete serenity the Congress. Of course I was Miranda. Carlotta was so good as not only solving problems but also preventing them. Running —actually flying—from side to side of the Lingotto, retrieving forgotten sheets and phones, coordinating my very fast changes of clothes and related accessories, entertaining guests… It was hard, but she did just fine. Those intense days have not undermined our friendship, indeed. So much so that Carlotta, who in the meantime has become a very good chemist, is the author of the nice drawings of this book.

The prayers of Grandma Severina

There's no event that I planned, from the simplest meeting with 20–30 participants to the World Congress with over 2,000 participants, that does not enjoy fantastic weather conditions. Grandma Severina who, unlike the rest of the family, prayed a lot, with prayer beads always in her hand, having noticed my great passion for organizing scientific events, before dying told me: "You can organize all the conferences you want, because I prayed so that they will always be a great success". And so it has always been. Sometimes, in the middle of a rainy day, the sky opens and the sun comes up, just for the duration of the event, which is amazing for those who does not know my secret—grandma's word. The only thing that Grandma Severina could not predict was the Covid pandemic!

Congresses in the pandemic era

The Covid-19 pandemic is changing our way of living, working, socializing. Also Congresses and Meetings are affected. Being impossible at the moment to people to get together in big numbers as well as travel long-distance, distance meetings, from zoom to webinar are becoming more popular. At the moment, they permit researchers to continue meeting, though virtually, and interacting. In the future, when the situation will normalize, such form of communication will probably remain, being effective in many situations. Hopefully, anyway, we will resume our old and pleasant attitude to meet personally!

Agroinnova, A Wonderful Adventure!

It would take a whole volume and not a simple "spore" to describe this beautiful adventure, but I will try not to get carried away too much by enthusiasm. Agroinnova is a Centre of Competence for the Innovation in the Agro-environmental field that I co-founded in 2002. Why? To be able to work at the speed of a private enterprise while remaining within the University. To be able to better use the many very relevant grants we were able to get from private and public bodies (see box page 112). To be more competitive at European and international level. To show that even the University system can react quickly and effectively to companies' demand. How? Copying, with the Italian creativity, the model of the Centres of Competence developed in Switzerland, which I had the chance to know well as an evaluator.

Resources, passion, interesting topics to deal with (see box page 112): there were all the ingredients for a new adventure. With a great contribution from the University governance, that helped us to build an agile structure, we have developed a Centre that is able to carry out at the same time basic and applied research, technology transfer, communication and lifelong learning. This is not the place to dwell on Agroinnova's activities. Rather, I would like to focus on the passion that animated the many people (see box page 113) who helped to make this small Centre a structure appreciated in the world.

All of them, in fact, in every area, have contributed with their commitment and expertise to the development and growth of the Centre, increasing its efficiency and ability to seize opportunities for collaboration, bringing significant resources to the University of Torino.

First of all, an effective and competent administrative office, able to best assist researchers in the management and administration of resources. Technical staff motivated and aware of the importance of their role, which consists in monitoring all the field trials carried out by Agroinnova. External collaborators, with specific skills, located in Italy and all over the world, able to follow with passion technology transfer activities, adapting to the habits of the most different countries. Young and motivated researchers able to appreciate the possibility of working in a stimulating environment, with modern facilities and up-to-date equipment.

Agroinnova has given space over the years to people full of initiative, eager to make experience and ready to move abroad to know different realities and acquire new skills.

After eighteen years, it can certainly be said that Agroinnova proved that even in our country, despite bureaucracy, you can be effective and efficient.

And the business world has perfectly grasped and appreciated the Centre's ability to deal quickly and decisively with the practical problems that the operating world is constantly faced with.

The most beautiful compliment received? Be mistaken for a private enterprise, which happened more than once (see box page 113).

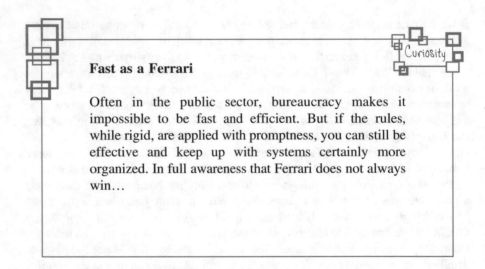

Fast as a Ferrari

Often in the public sector, bureaucracy makes it impossible to be fast and efficient. But if the rules, while rigid, are applied with promptness, you can still be effective and keep up with systems certainly more organized. In full awareness that Ferrari does not always win…

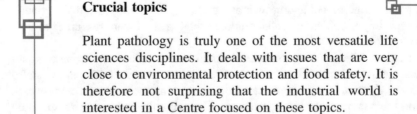

Crucial topics

Plant pathology is truly one of the most versatile life sciences disciplines. It deals with issues that are very close to environmental protection and food safety. It is therefore not surprising that the industrial world is interested in a Centre focused on these topics.

A great team

Agroinnova's success in these eighteen years is due to the dedication of those who have been working with passion: Manuela, Renata, Gianna, Domenico, Guido, since the beginning, and then, Gabriele, Ileana, Paola, Grazia, Andrea, Giulia. Everyone, working seriously in their own field, has contributed to the growth of the Centre. Always with team spirit and much passion.

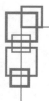

Curiosity

Agroinnova: Academic soul, company muscles

A journalist and friend, probably due to his longtime acquaintance with Agroinnova, developed one of the most effective definition for the Center: academic soul, company's muscles. To best explain our attitude to try to work with a speed of an enterprise, still being a public body.

Little Is Beautiful: From Saluzzo to Grugliasco

It will be the millions of miles accumulated with my continuous travels (as in the movie *Up in the air*, with George Cloney, do you remember?), it will be the beauty of the Grugliasco Campus and the big commitment in Agroinnova, but certainly "landing" at Grugliasco, an industrial town not far from Torino, where we moved in 1996, has helped to slow down a bit (so to speak) my natural vocation to travel. Or better, it brought me back to the calm of Saluzzo—while making the necessary comparisons. I would not want to offend my fellow citizens of Saluzzo, since nothing of Grugliasco resemble the beauty of my hometown, but its size, the same atmosphere of provincial town, the green of the campus have probably brought me back to feelings and emotions I experienced in my early years in Saluzzo. Not to forget that the last stretch of road that takes me from home to Grugliasco allows me to admire, every morning, my beloved Monviso.

In Grugliasco there is everything I need: bright and well-equipped laboratories, huge greenhouses, skilled technicians, an administration open to collaboration (see box page 116), interesting companies to work with…

So why, after so much running, do not stop for a while and try to make use of what learned around the world in favour of this territory?

This question, which I have been asking myself for some years, has finally been answered with a conscious mind. So my many suitcases spend more time resting, finally enjoying, even them, a little respite.

Less stressed than I was in the past when I was travelling constantly around the globe, I enjoy much more the beauty of the campus and its modern facilities, feeling the great satisfaction that teaching and contacts with students can give, together with the effort (although pleasant) of directing a Centre like Agroinnova. With the same passion I had as a scholarship student, but enriched by many years of experience, I can now devote completely to research and writing, with more time to spend with schools (see box page 116), as well as helping my hometown in a new adventure (see box page 117).

M. L. Gullino, *Spores*,
https://doi.org/10.1007/978-3-030-69995-6_26

A very receptive territory

A strong link with the Grugliasco administration since the beginning permitted Agroinnova to engage with the territory in a lot of activities, such as visiting and hosting schools, meeting teachers, organizing games for schools. A very intensive and fruitful cooperation indeed.

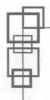

Working with Schools

Grugliasco, with its 38,000 inhabitants and a very large and efficient scholar district, provided Agroinnova the opportunity to work with teachers and children of kinder gardens and primary schools. Hosting kids in the campus and organizing activities in their own classroom. A very enthusiastic experience! With our young researchers and Ph.D students strongly involved in the activities.

Gifts from life

In December 2020 the municipality of Saluzzo launched its bid to the Italian Capital of Culture in 2024. With a very glorious past in its history, having been the Capital of the Saluzzo Marquisate in the Middle Age, Saluzzo is ready indeed for a new challenge. A wonderful gift to me to be able to help, in my role of Saluzzo-borne person, bringing this town in my hearth, as well as because of my role at the University of Torino, as Vice-Rector for the valorisation of culture. A new adventure is starting.

The Secret World in the Soil

After spending some years to study fungi that cause diseases affecting the aerial part of the plants (the so-called foliar diseases), one day I was asked to start a research on soil-borne fungal pathogens. The Institute of Plant Pathology where I did spend most of my career, except the years spent abroad, boasts a long tradition of research on some soil-borne fungi, especially *Fusarium* and *Verticillium* species, responsible of vascular diseases in a large number of plants. By colonizing the woody vessels of the affected plants, these fungal pathogens plug the passage of water, causing wilting and premature death.

The soil and its hidden world—an incredible number of microorganisms—represented for me a great discovery. There are microorganisms useful for the development of plants, such as nitrogen-fixing bacteria, microorganisms that in themselves have a very neutral role, since apparently they do neither good or bad, pathogens (our object of study as plant pathologists) and any antagonistic agents.

Proportionally, from the numerical point of view, the pathogens are often a small minority, but their aggressiveness and the ability to cause damage make them very dangerous: few but good, in short. Or rather, few and bad.

Different is the case of soil-borne pathogens: some are able to invade the soil and, having a so-called "resistant" structure, which allow them to survive for a long time and in great number, they remain mostly indifferent to many soil treatments. As their density increases, plant damage tends to increase. These characteristics are typical of fungi like *Fusarium* and *Verticillium*, which have attracted the interest of many scientists, precisely because of the serious damage they can cause to crops of great economic value. The genus *Fusarium* has differentiated in the time hundreds of so-called special forms, able to attack specific hosts. In other words, there is a *Fusarium* able to infect tomato, one specialized on carnation one, another on basil and so on. You can say that almost every plant has its own *Fusarium*. Different is the case of *Verticillium*, which is a genus with "a good appetite" (the right term is polyphagous), attacking many different hosts.

Moving from the soil towards the surface of the ground, we find other fungal pathogens, such as *Rhizoctonia*, *Sclerotinia*, *Sclerotium* which also attack various

M. L. Gullino, *Spores*,
https://doi.org/10.1007/978-3-030-69995-6_27

host plants—and the latter two are able to produce resistance structures (sclerotia) making them live longer, even in the absence of the host plant.

A common characteristic of all these pathogens is their tendency to proliferate in absence of crop rotation. Intensive cultivation systems, where the same crops (e.g. vegetable or flower crops) are grown in the same place, over limited surfaces, for many years in a row, are likely attacked by selected pests. Hence the need to counter the increase of these pathogens in the soil, by various means, often very drastic, developed over the years.

Fortunately in the soil there are also "good" microorganisms, that is, fungal and bacterial pathogens able to control the development of some plant diseases to a lesser or greater extent (see box page here below).

Furthermore, soil contains not only plant pathogens but also animal microorganisms and, above all, plant roots that with their exudates can attract or repel pathogens. It is therefore a complex hidden world, very hard to study, so much so at times a soilless system is needed in order to avoid soil complexity (see box page 121). Nowadays, modern techniques permit to exploit was is called the soil microbiome, better understanding the deep relationships existing among the different soil-inhabitants.

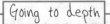

Going to depth

Suppressive soils

In some soils, due to their chemical-physical structure and/or the presence of antagonistic micro-organisms, certain soil-borne pathogens—although present and despite the cultivation of crops susceptible to them and the presence of favourable environmental conditions— do not cause the appearance of the disease that everyone would expect. We are talking of suppressive soils.

This phenomenon, known since the 1800s, has been observed in different situations, although always very particular and sporadic, and has been studied in several countries. Also in Italy we have our suppressive soils, observed in Albenga and studied in the years 1980s and 1990s by the plant pathology group of Torino. In these particular soils, Ligurian farmers did not observe Fusarium wilt on tomato and carnation; this was due to the presence of antagonistic *Fusarium* able to compete with pathogens for niches and essential nutrients.

Unfortunately, there are few lucky farmers. However, the isolation of the "good" *Fusarium* from such suppressive soils has allowed to exploit this phenomenon developing biocontrol agents to be applied against soil-borne diseases.

Soil, no thanks!

In the last fifty years, so-called "soilless" cultivation, carried out replacing the soil with different substrates (rockwool, perlite, peat...) or even simply nutrient solutions, seemed to be a promising solution against soil-borne diseases. This type of cultivation, developed in the United States and in Northern Europe, and now widespread also in the Mediterranean countries for high-income horticultural crops, allows to reach very high levels of production, avoiding soil overexploitation. These cultivation systems, in fact, provide for the recovery of the circulating solution, thus avoiding soil contamination.

Although apparently perfect, born to favour the highest yields, and considered almost aseptic, also soilless cultivation systems are not free of phytopathological problems. In fact, if the plant material (seeds, cuttings etc.) gets infected by pathogens, the latter can spread with great ease and speed in the substrates and in the nutrient solutions, with devastating effects on plants. That is why plant pathologists have started a new research field: "disease management of soilless crops".

Solar Energy for Weakening Soil-Borne Pathogens

Very often the most interesting results in the agricultural sector come from acute observations by farmers. It was the mid-1970s when an Israeli farmer noticed that the fields in which plastic films, used to reduce weed growth during the summer, were kept on the ground longer presented, subsequently, more limited attacks than some soil-borne fungal pathogens (the same descripted in the last chapter). The farmer talked about it with a young researcher named Avi Grinstein and a famous professor, Jaacov Katan. From their elaboration of the farmer's observations, came out the solarization, a very innovative technique of soil disinfestation that exploits solar energy.

Solarization has been considered among the 100 agricultural innovations of 20th century. But what is it about? It is a method of soil disinfestation through solar energy (and therefore a climate-dependent process, which can be used more effectively in warmer countries) aimed at raising the temperature of the first 30–40 cm of the soil in order to reduce the development of soil-borne pathogens. During the warmest and best insolated periods of the year (from June to August in Italy), soil must be covered for at least 4 weeks with transparent plastic films, after being irrigated abundantly, so that the water favour the transmission of heat in the soil from the surface in depth. In fact, exploiting solar energy, the soil undergoes a real pasteurization, reaching, in the most superficial layers, 42–50 °C, a temperature capable of eliminating, or at least devitalizing mycelium, spores, resistance structures of many soil-borne pathogens—not only fungi, but also nematodes, insects and weed seeds.

How does solarization work? With a double mechanism, physical and biological. The thermal effect eliminates or, more often, weakens the pathogens, which are so more easily subject to the antagonistic action of some microorganisms. Solarization has a considerable and undisputed value: it can be practiced with reduced costs and has a low environmental impact.

Avi Grinstein and Jaacov Katan have become world famous (see box page 124) and solarization is now used wherever the environmental conditions allow it. To date, this method, which has been rated among the 100 best invention in agriculture

© The Author(s), under exclusive license to Springer Nature Switzerland AG 2021   123
M. L. Gullino, *Spores*,
https://doi.org/10.1007/978-3-030-69995-6_28

of the XXth Century, has been tested in about 80 countries, including the grey and
rainy Great Britain. Of course, it works better in the warmer countries, such as the
Mediterranean ones (see boxes page 124). In Italy it is also applied under greenhouses (see box page 125).

Curiosity

Women's tights useful for research

Avi Grinstein and Jaacov Katan have worked together for
years, with great passion and dedication, until a sudden
death took Avi away. I wish to remember a curious
episode about them. In order to study solarization effect
on soil-borne pathogens, they used to bury in the ground
the survival structures (sclerotia, chlamydospore, etc.) of
different fungi at various depths, enfolded inside bags.
Over time, Avi and Jaacov had identified women's tights
as the most suitable fabric to enfold fungal sclerotia and
mycelia. I let you imagine the astonishment of an Israeli
colleague when he opened the door of Prof. Katan's
office and saw him near the window with Avi, comparing
at the light the thickness and transparency of various
samples of women stockings...

Eris, a great Greek friend

Solarization works very well in all warm countries.
In Greece it has been studied by a wonderful friend
and colleague, named Eris Tjamos. Eris, Professor at
the University of Athens, is not only a competent
colleague, but also a wonderful friend. Everybody
knows his hospitality. Eris did organize many
Congresses and Workshops, in the most beautiful
locations in Greece, islands (Crete, Corfu, etc.)
included.

A longterm cooperation

When I began my adventure as a plant pathologist, my supervisors warned me: "Never start to work on a subject already studied by Israeli colleagues, unless you want to work day and night". In fact, by working closely with many Israeli researchers, I have clearly understood the meaning of that advice.

The collaboration between the plant pathology group of Torino and Israeli colleagues, started in the 1960s, and is still very intense: we are now in the fourth generation of researchers who keep on interacting with commitment and imagination.

Going to depth

What about solarization in Italy?

In Northern Italy, where insolation is not always enough to allow this practice, Angelo Garibaldi, at the former Institute of Plant Pathology of the University of Torino has developed the use of solarization in greenhouses, thus combining the effect of insolation with the greenhouse effect. In Central and, above all, Southern Italy, climate conditions allow to successfully use solarization in open field.

My First Spin-off Out of Rotten Apples

Since I was a child, I have learned to be quite independent, trying to ask my parents as little as possible. At the same time, I have always had a good sense of business, or, at least, I think so. That is way my father offered me, when I was 12–13 years old, to sell rotten apples for obtaining profits that I could freely dispose of.

It was normal, at that time, to sell scrap apples (small, partially rotten) to ciders. This trade, although not very simple from the logistic point of view (you have to organize transports, deliveries…), was very interesting to me, because it allowed me to satisfy my wishes as a girl (the first scooter, 45 laps, shoes, books etc.) without burdening my parents. This was my first spin-off!

Certainly at that time I was totally unconscious of the risks of providing agro-food industry with not excellent material.

Later on, in the early 1970s, attending the course of Ecology at University, I became aware of the possible contamination of cider and apple juice by patulin, a dangerous mycotoxin of food-safety concern produced by the fungus *Penicillium expansum*, responsible for the post-harvest rot of apples.

I was deeply shaken. That night, when I got home, I did some research. At that time patulin was a mycotoxin still little studied and little was known about its possible toxicity to humans and animals. But it certainly could harm the health of consumers. The mere thought of having bought my scooter speculating—although without knowing—on the health of consumers, made me feel totally unease.

The first reaction was not to get on that scooter anymore (with great joy of my brother who immediately took it for himself). The second, certainly more rational, was to decide to start—in the future—carrying out research on mycotoxins.

And so, finally, in the 1990s, the right time came to start some research on the possible contamination of fruit juices by patulin.

In the meantime, knowledge about the toxicity of this mycotoxin, which affects various products of the horticultural sector, had been increased, having been found in many fruits, especially apples and pears, but also grapes, peaches, apricots, cherries, plums, bananas, strawberries, melons, small red fruits and their unfermented derivatives. *Penicillium expansum*, the main patulin fungus, easily

M. L. Gullino, *Spores*,
https://doi.org/10.1007/978-3-030-69995-6_29

penetrates the fruits in their final stage of conservation, through wounds or natural openings, causing the appearance of a typical green-blue rot that everybody, at least once in life, have the chance to observe.

Patulin was isolated for the first time in 1942 by American researchers and over the years several studies have been carried out to evaluate its toxic properties. These studies have shown that the intake of patulin with contaminated products causes damage mainly to respiratory and digestive tract. This mycotoxin, classified by the International Agency of Cancer Research (IARC) as "non-cancerous to humans and possible to animals", is potentially harmful to consumers of products derived from contaminated fruits, for example juice and stewed fruit.

Researches carried out with a team of collaborators to assess the presence of patulin in fruit juices marketed in Italy have shown that it is the lower quality juices that are most frequently contaminated by patulin, although, fortunately, at generally low levels. A single sample of the hundreds of fruit juices tested showed levels of patulin above the legal limits. The European legislation, as well as that of most industrialized countries, sets the maximum permitted amount of patulin and other mycotoxins. Higher quality juices and stewed fruit were not found in our studies to be contaminated with dangerous levels of patulin.

Measures to prevent the presence of patulin in processed products include the use of healthy fruit for their preparation.

So the research I have been doing has not reduced my guilty feeling, but actually increased it. Anyway, consumers have not to worry, since today products are strictly controlled. Special attention goes to baby food (see box page here below). For sure, thinking of my very first spin-off, I can say that plant diseases were in my destiny (see box page 131).

Curiosity

Baby food

It will be of interest to the consumer to know that national and international legislation pays special attention to baby food. This is because on the one hand children are more susceptible than adults to the possible consumption of products with some toxicity and, on the other hand, because in proportion, the amount of food ingested in the first months of life, compared to weight, is greater in infants than in adults.

From rotten apples to Agroinnova, the golden apple

A few months ago, a colleague, during the meeting of the Scientific Committee of Agroinnova, maybe considering the change of leadership that will take place in a few years, defined the Center a "golden apple". Though the intention was probably not exactly friendly, I took it, as I always do, in a positive way. I have gone indeed a long way! From the rotten apples of my first spin-off to the golden apple.

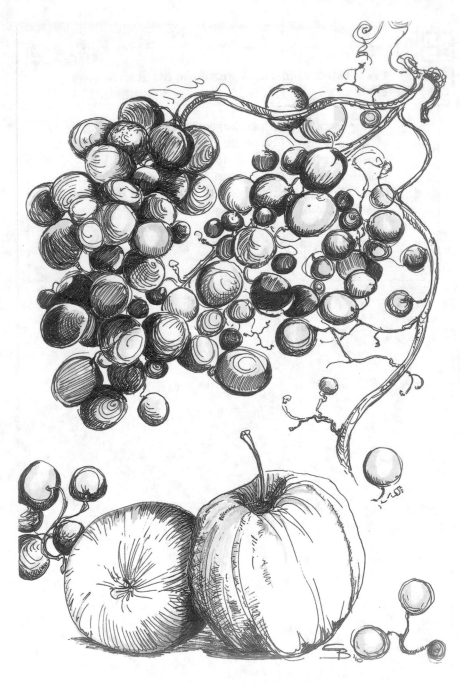

Looking back to my past

Nowadays, after a long-time experience in research, and some experience with real spin-offs, I am not ashamed anymore about my first business with rotten apples. After all, that's was my first, very early spin-off. A good training for learning about business. Also learning to be economically independent, something very common in the United States, but not so much in Italy, where children often grow up much spoiled, with very limited experience in earning money.

Some Promotion of Italian Wines

In recent years, researchers focused their attention on the evaluation of possible wine contamination from ochratoxin A (OTA), a mycotoxin produced by various *Aspergillus* and *Penicillium* species, which usually colonized wine grapes around harvest time.

Several international projects have shown that in fact, some wines, especially those produced in the areas with the warmest average temperatures during the summer (such as Southern Europe), are contaminated by OTA levels which are sometimes higher than those considered dangerous by European legislation.

Ochratoxin A represents a serious threat to human and animal health, both directly or indirectly. Several animal studies indicate that OTA may perform hepatotoxic, immunosuppressive, teratogenic and carcinogenic activity depending on the dose and species concerned. The toxin builds up in blood, liver and kidneys.

Considering the human exposure to OTA and the abundance of experiment data on animals, the World Health Organization and the Scientific Committee on Food have set the safe level of exposure to OTA below 5 ng/kg body weight per day. Some European countries have proposed technical barriers setting different rules for the maximum safe concentration of OTA depending on the country.

The first study on OTA contamination on wine dates back to 1996. Since then, OTA's presence in both local and imported wine production has been reported in many European and non-European countries (see box page 134).

High levels of contamination have been found in southern European red wines (for example Greek, Portuguese wines, and Italian wines produced in the south), which are usually more contaminated than white ones from the same wine-growing areas. Wine is not the only grape derivative in which OTA was reported: in fact, mycotoxin was also found in juices and, at much higher concentrations, in raisins.

The contribution of wine to the daily tolerable OTA intake may be considered negligible in the case of North European wine drinkers, but this does not apply to average drinkers who consistently consume red wine produced in southern Europe. In this case, wine alone can contribute to the diet an equal or even higher OTA amount than the maximum safe concentration indicated by the World Health Organization.

Since 2006, limits for the presence of this mycotoxin in wine and grape juice have been set to 2 ppb (ng/ml, i.e., nanograms/ml).

Luckily OTA is not present in most Italian wines (see box page 135), so we can keep on drinking them—in moderation, of course! (see box page 136).

Lady OTA

Several European projects on OTA in wine have been coordinated by Paola Battilani, professor at the University of Piacenza, who caught the urgency of this topic for European and Mediterranean viticulture in particular. Sporty and very efficient, Paola has been able to lead European groups that very timely drew the attention of stakeholders on this issue, also providing legislators with useful information to introduce limits to OTA levels in wines.

Going to depth

The situation of Italian wines

Studies on Italian wines have shown that in the vast majority of the samples analysed the concentration of ochratoxin A did not exceed 2 ppb (thus, 2 part/billions, which corresponds to 2 g/l, limit provided by law). In wines from northern Italy (that is Piedmont, Liguria and Emilia Romagna), the average concentrations of OTA are extremely low, as has been shown by many researches. The situation is more critical for wines produced in the central-southern part of the Italian peninsula, where OTA values are five times higher than those observed in the northern regions. Some samples, however, showed contamination levels below the legal limit: it occurred for wines produced in southern Italy.

In the samples analysed OTA concentration was quite similar in red and white wines. Contamination by OTA was higher in wines from the southernmost parts of Italy than in the northernmost ones.

It therefore seems advisable to limit OTA content in southern wines by adopting preventive strategies already in the vineyard, during the most dangerous phenological phases—veraison and harvest. A Piedmontese person like me would say: go with Barolo and Barbaresco! In the meantime, it must be considered that climate change, with increased temperatures, might cause, in the future, higher contamination by OTA of wines, since *A. carbonarius* growth is favoured by higher temperatures. Another example of how climate change may affect agriculture.

How much do we risk drinking Italian wines?

Some researchers of the Italian National Institute of Health evaluated the exposure of the Italian population to ochratoxin A, connected to wine consumption. 1,166 samples were analysed over the years 1988–2004, on wine vintages of 19 Italian regions: 773 red wines, 290 white wines, 75 rosé and 28 dessert wines.

The results showed that a sample of red wine was the most contaminated, with a concentration of OTA residue of 7.50 ng/ml, although rosé wines had the highest average concentration of OTA. In addition, the study reported an increase in OTA contamination from North to South.

The contribution of wine to the exposure to OTA of people has also been calculated on the basis of the estimated wine consumption in two different databases, and also on the basis of the worst-case scenario, that is when a higher wine consumption is expected, but in both cases OTA exposure was not a concern.

However, it should be recalled that this study only addressed the risks connected to OTA exposure in wine consumption. The fact that the risk is low with Italian wines does not entitle us to overdrink.

A Pocketful of Chestnuts

A pocketful of chestnuts was the title of an Italian comedy directed by Pietro Germi. Good and healthy chestnuts have to make it through several pest attacks—and not just plant pathogens. Also for this reason, during the past few years, in Italy, which is one of the main producer and consumer, chestnuts have become rare and expensive. Perhaps the most famous pathogen is the cortex cancer agent, *Cryphonectria parasitica*, first found in 1904 in New York, from where it quickly spread to North America. By the end of 1920 it had reached the Pacific coast, where it destroyed the American chestnut forests. In Europe, the disease was reported in Belgium in 1924, in Germany in 1927 and in 1938 in Italy, where it found ideal conditions in the chestnut groves of the Apennines and the Alps, before spreading to many other European countries and regions of Asia.

Affected plants show rust or blackened coloured lesions on stem and branches and the typical necrosis of bark tissues (bleeding canker) in case of fast attacks. In the case of slow attacks, however, the plant can recover. The pathogen produces toxins capable of destroying tissues at a distance, as insects that live in the bark and birds help to spread its spores.

Management strategies of chestnut canker include interesting biological control methods, such as the use of hypovirulent pathogen strains to make a sticky paste to apply on bark lesions. This biological control method was developed after observing that, in the field, in some cases, the plants showed very attenuated symptoms or, even, presented recovery signs. This was due to the spread of attenuated strains of the pathogen from one plant to another, through contact at root level. A sort of vaccination against the disease.

Recently chestnuts have been seriously affected by a fungal pathogen belonging to the genus *Gnomoniopsis*, which has caused severe yield losses, especially in north-western Italy (see box page 138). The severity of this disease is such as to jeopardize the production of the famous marron-glacé, whose price has greatly increased.

But chestnuts' issues do not end here, as the cases of aflatoxin contaminated flours are more and more frequent, and a very dangerous insect is devastating chestnut cultivation in Italy (see box page 139).

© The Author(s), under exclusive license to Springer Nature Switzerland AG 2021
M. L. Gullino, *Spores*,
https://doi.org/10.1007/978-3-030-69995-6_31

137

In short, growing chestnut trees, which for their rusticity were in the past the ideal crop for the marginal areas of Italian mountains, is increasingly difficult.

In the absence of natural antagonists in Europe and in the obvious impossibility of using chemical control means, it is crucial to monitor and follow the evolution of infestations and implement counteracting strategies such as biological control. The entomophagous insect *Torymus sinensis*, native to Japan, is a parasite of chestnut gall wasp larvae. Agricultural entomologists of the University of Torino were able to reproduce this natural antagonist in large quantities so to introduce it in all those areas where Asian chestnut gall wasp attacks are more serious. Such approach has been very successful.

Curiosity

Nucleic magnetic resonance (NMR) for internal quality evaluation of chestnuts

Chestnuts are well suited to new diagnostic methods, which involve the use of magnetic resonance imaging, a technique used for clinical diagnosis since 1980s. As a non-invasive technique, it may have very interesting applications in the agri-food sector for quality evaluation of some organoleptic characteristics of fruit and vegetables. Why NMR is currently highly evaluated for chestnuts analysis? Because chestnut is a product of high economic value and, due to its dimension, it can be easily manipulated with such technique.

Going to depth

Asian chestnut gall wasp: an alien attack on European crops

A chestnut pest, imported from China a few years ago, represents such an interesting case that it takes me, for a few lines, beyond my strict area of expertise. Since 2002, in fact, chestnut tree has been affected by more and more worrisome infestations of *Dryocosmus kuryphilus* known also as "Asian chestnut gall wasp", native to China and introduced accidentally with propagation material coming from the Far East or the United States. This insect is currently the most serious threat to chestnut crops in Italy. It causes the appearance of various green and reddish galls on leaves, buds and catkins, disrupting the fruiting process and then reducing yield.

The Incredible Success of Ready-to-Eat Salads

Who of us, busy women always in a hurry, is not tempted by buying bagged salads? Smart, attractive, ready-to-eat. And we do well, because behind this product there is so much work, so much technology and innovation!

Italy is one of the leading countries in the production and consumption of ready-to-eat salads and this is among the horticultural products that has been least affected by the recent economic crisis and the related decrease in demand. The increasing appreciation of fresh-cut vegetables—term referred to any minimally processed products sold already washed and packed—has led to an intensification of cultivation in many geographical areas, with 5 to 6 production cycles per year on the same soil.

The intensive nature of this cultivation system is connected to the development and spread of several phytopathological problems, caused by the emergence of new pathogens which sometimes, if not effectively managed, may cause very serious yield losses (see box page 142).

Studies carried out since 2002 have made it possible, for example, to observe only in Lumbardy (northern Italy), where leafy vegetables for the ready-to-eat sector are very intensively grown, the continuous appearance of a significant number of new pathogens, mostly soil-borne fungi, threatening production of lettuce, rocket, lamb's lettuce, romaine, etc. (see box page 143).

Many of these pathogens spread through infected seeds. Seeds are in fact among the most effective means of long-distance transport of pathogens, especially when they are produced in a few specialized facilities to be then marketed worldwide. This trend in the production of propagation material—widespread in the case of leafy vegetables—on the one hand, has led to the seed quality improvement, thanks to the availability of very sophisticated technologies, but, on the other hand, has favoured the rapid spread of pathogens previously unknown in many geographical areas.

A further issue to consider in relation to leafy vegetables production for ready-to-eat sector, is consumer protection, with special regard to both the absence

© The Author(s), under exclusive license to Springer Nature Switzerland AG 2021
M. L. Gullino, *Spores*,
https://doi.org/10.1007/978-3-030-69995-6_32

of agrochemical residues in foodstuffs, and the possible contamination by human pathogens, such as *Salmonella* species or *Escherichia coli*.

Therefore, that of fresh-cut salads is not only an expanding sector but also a sector that requires, due to the complex plant-health situation, in-depth and innovative research. It is in fact important to adopt preventative control measures in order to avoid the presence of pesticide residues in ready-to-eat products.

For the technologies involved in growing, processing and marketing fresh-cut vegetables, the ready-to-eat sector represents a challenging research field and an interesting topic for the development of new sophisticated and innovative pest control methods.

Going to depth

Intensification of crops and emergence of new plant diseases: rocket as an example

Rocket, both cultivated and wild, has become a very interesting species, also because it is an important component of ready-to-eat salads. In recent years the intensification and spread of rocket crops has caused the appearance of several pathogens never seen before.

Many of these pathogens are transmitted through infected seeds: a very low percentage of contaminated propagation material is sufficient to cause serious damage to crops. Phytosanitary control is therefore extremely important as a prevention measure, effectively implemented through sophisticated diagnostics, in order to use good quality seeds or, if necessary, treated ones in order to decontaminate them.

Gianna, the fairy of the ready-to-eat salads

Since 2002, when she started working at Agroinnova on diseases of leafy vegetables for the ready-to-eat sector, Giovanna Gilardi, visiting hundreds of greenhouses and tunnels throughout Italy and beyond, with the nose of a truffle hunter, has discovered at least fifty new diseases, previously unknown in Italy and, in some cases, even in Europe and worldwide. There is in fact no new symptom, any particular crop situation, which escapes her attentive and expert observation.

Gianna has become the fairy of the fresh-cut salads: she does not only focuses on new diseases detection, but also works on sustainable disease management of leafy vegetables, and has become an important reference for growers and extension services. For this reason, Italian farmers specialized in salads production for the ready-to-eat sector immediately contact her when they are in trouble with some apparently new or unknown pathogens.

Falling Trees

Trees significantly contribute to the beauty of our cities, even though certainly they do not have easy life in such environment (see box page 147). Despite such a situation, the Covid pandemic showed us the resilience of nature, including urban trees (see box page 148). The management of the tree heritage can be a real challenge for our communities (see box page 148).

How many times have you heard about trees, even large ones, seemingly in good health, suddenly crashing? And not just in the presence of adverse weather events.

The stability of ornamental and monumental trees may be compromised by the presence of root rot or decay of the central cylinder of the basal parts of the stem. Decay process in wood structure is caused by fungi, often of Basidiomycota group. But how can you tell if a tree is healthy or at risk?

The development of a sporocarp of a lignicolous fungus, that is able to destroy wood, on a tree indicates that the plant itself is affected by a disease potentially able to reduce mechanical strength of branches and stems. Although the only presence of a fruiting body does not allow an evaluation of the extent of the wood decay, it may allow the diagnosis, that is, the identification of the genus and the species of the fungus that infected the plant.

To understand the extent of wood decay, an instrumental survey provided by the so-called Visual Tree Assessment (VTA) involves, in addition to visual inspection, the use of special inspection technologies to evaluate the risk of the tree falling. These technologies, combined with visual analysis, allow a quantitative estimation of possible damages that may compromise trunk stability. VTA consists of three phases: visual inspection of plant vitality and external damages, instrumental examination of any damages found and, finally, evaluation of the mechanical characteristics of the healthy residual wood (see box page 149).

In recent years, minimally-invasive procedures have been developed to assess the likelihood of whole tree failure, such as tomography. Developed initially for medical purpose, these techniques have been later extended to other sectors, thanks to the evolution of measuring and calculations instruments.

M. L. Gullino, *Spores*,
https://doi.org/10.1007/978-3-030-69995-6_33

The internal damages detected by these instruments and the analysis described are represented by wood-decay, that is, the degradation of wood and plant material caused by fungi, which may lead, in extreme cases, to the development of cavities in the stem. Quantitative analyses have shown that an internal cavity affecting more than 60% of the trunk diameter significantly increases the stress on the residual part of the trunk, which has been also confirmed, in almost all cases, by observations of fallen trees presenting these phenomena.

Although almost twenty-five years have passed since the first applications, the issue of assessing tree stability in the urban environment is always very topical and the presence of good specialists is crucial.

More generally, to manage the tree heritage of our cities, trained and sufficiently experienced technicians are needed and it is important for public administrations to understand that you have to invest in the maintenance of the green heritage. Unfortunately, despite the attention paid and the inspections carried out, trees falling in avenues, parks and gardens is not so unusual, with damage to things and/or people and consequent judicial proceedings (see box page 150).

An important aspect which all of us, as citizens, have to understand, is that trees also have a life cycle and that it is often absurd to persist in keeping decaying trees alive. Trees, in fact, can always be removed, when severely affected by diseases, and replaced (see box page 151).

Urban trees' hard life

It is certainly not easy for trees to adapt to urban life: sunlight is reduced by smog, rainwater drags acid substances with it, asphalt certainly does not facilitate soil aeration and root development, pavements prevent water from filtering, city's soil is very poor of nutrients and crossed by pipes and pipelines. And in addition to all this, there is human carelessness for the environment, including plants.

Trees really struggle for survival in the cities. Growing old, they show more and more symptoms of site disturbance and lapse into a progressive decline. Public administrations, with increasingly limited budget, tend to reduce the resources allocated to green heritage management. In Italy, Milan invests, for its urban green care, 1.1 Euro/m^2, Torino and Florence, 0.80, Rome 0.40. Unfortunately, the negligence of public authorities towards the environment (trees do not vote, after all) often leads to cut off more funding in this sector than in others.

When you cut resources to cultural activities, people complain, while few ever notice administration's cuts to green management and protection, until maybe, as it already happened, a falling tree causes a fatal incident. Then who is responsible? Nobody wants to take charge of it. Unfortunately in some cities, including Torino, it took some accidents to draw the attention of the authorities on this urgent matter.

Trees in the post-Covid era

During the lockdown imposed by the Covid-19 pandemia, also the less environmentally careful people, could not avoid to notice that, while we were closed in our houses, plants did continue to grow. During the spring 2020, throughout the windows, we did observe trees leafing, flowering, as nothing happened. Hopefully, the experience that we lived will make us much more environmentally respectful. We should have understood how resilient nature can be…

Torino, a green city

The city of Torino, with its magnificent avenues, parks and gardens, boasts a tree heritage of 110,000 specimens, to which must be added the 50,000 trees on the hill side. The arboreal species classified in Torino are more than 70, whose most represented are plane (15,000 specimens), linden (10,000), hackberry and maple (5,000 each), and horse chestnut (4,000). The amount of urban green space per person is about 22 m^2.

Torino is, by tradition, a city that devotes great attention to its green heritage in general, thanks to the careful and efficient green maintenance carried out by a historically very well trained team of experts.

Going to depth

Tree stability evaluation (TSE)

The purpose of the TSE method is to assess the probability of a tree falling; this does not imply the calculation of a precise hazard value, nor does it mean accurately predicting when the tree will fall. Rather, the method attributes the inspected plant to a certain phytostatic risk class. Each risk class must be statistically interpreted in terms of event prediction. Indeed, even a completely healthy and flawless tree can fall in case of exceptional natural events. TSE's strength is the consideration of both biological and mechanical aspects for the assessment of tree stability.

Mechanical aspects estimation requires instrumental surveys, performed with resistograph type penetrometers and electronic hammer. Resistrograph, by far the most used and accurate in TSE analysis, is a sophisticated electronically controlled drill, equipped with a probe of variable length, which advances into the tree at constant speed, adjustable according to wood density. The energy consumption during drilling, graphically visualized by a dendrogram, measures the mechanical quality of wood. The decayed wood, opposing a lower resistance to perforation, induces a dendrogram lowering.

The electronic hammer, instead, measures the speed of propagation in the wood of sound waves generated manually by tapping on a pin inserted in the wood. The speed values of the sound wave are compared with those reported in the literature for the various tree species, obviously healthy.

The collaboration with Torino Public Prosecutor Raffaele Guariniello

You may wonder what ever Prosecutor Raffaele Guariniello has to do with trees...Very famous worldwide for his work on asbestos, Raffaele Guariniello also dealt very effectively with the problems of falling trees. It is precisely because of a tree fell in a golf club in Torino in 2008 that I got to know him personally. After having long wished the intervention of this official—renowned for his intransigence, preparation, environmental and safety attentiveness—to solve the many unpleasant situations we come across every day (gym changing rooms far from perfect, expired food in supermarkets...), at last I had the opportunity to meet Dr. Guariniello (although unfortunately due to a serious accident that caused the death of a golf player) and collaborate with him in several cases, always about trees. It can happen to be disappointed by these encounters in vivo, as we say in scientific jargon. This was not the case: besides being a great magistrate, Dr. Guariniello is in fact a very attentive, kind, witty and nice person, able to go straight to the point on any issue. He has made people safety and environmental protection his reason for living.

To a researcher like me could certainly not escape the purpose of his actions, always aimed at making the best even of the most negative situations, so to improve knowledge and not repeat past mistakes. One example among many: the great asbestos trial. The Observatory he has set up to record the deaths from mesothelioma will allow in future to avoid other dramatic situations as those occurred in Casale's district. I was also very impressed by the effective team of collaborators he maneged to gather over the years. And if, before I had the chance to meet him, I thought it would take 100 Guariniello prosecutors to fix so many things that go wrong,

knowing him personally I realized that it would take a thousand, ten thousand people like him, scattered around the world, particularly in countries where asbestos is still used, despite the fact that nowadays its negative effect on health and safety is well known.

Loving trees may also mean having to cut them down

Trees, essential elements on an aesthetic and ecological level, become, unfortunately, often object of an obsessive defence, which does not take into account the fact that our urban trees are not always valuable specimens but rather nursery grown trees. So we must not wait for them to fall, but avoid the risk their roots blow up the road. Loving trees means to take care of them and replace them if necessary.

Trees that must be safeguarded and not removed are those that belong to natural reserves, to very rare biotypes and those that represent real natural monuments. All others must be respected, cared for, loved in a non-obsessive way. As well-aware citizens, concerned with environmental issues, we do not have to be surprised if a tree in a poor condition is felled to avoid serious accidents. Rather, we must demand its replacement.

The Neighbour's Grass Is Always Greener

The greater availability of free time, the desire for beauty and a healthy life in the open air, as well as the greater ease of travel and tourist trips, have increased the interest in turfgrass. For this reason, the attention paid to healthy and uniform grass is constantly and progressively increasing, both in the case of sport fields, due to high pressure of the media, and for ornamental and hobbystic purposes. Developed as a real crop in Great Britain, where climate conditions are extremely favourable, turfs are currently widespread even in geographical areas where the environment is rather adverse to their growth (box page 158).

Of course also turfs can be affected by diseases, especially those of golf courses and football playground, where high maintenance, together with requirement for perfection, favour stress conditions.

In fact, if we consider the turf of a green, in a golf course, we realize to deal with a real crop, subjected to numerous cuts, often excessive fertilization. Not to mention, until a few years ago, the incredible number of chemical treatments carried out by technicians, often without a specific expertise.

In the case of golf courses, several pathogens, especially fungi, can cause serious attacks and damage. The most famous and widespread disease of golf courses is called "dollar spot", whose name comes from the appearance of small circular spots, the size of the old American metallic dollar coin. In football fields, both fungal pathogens and stress situations, caused by trampling, may cause very serious damage.

Control strategies have always been particularly difficult in such conditions, due to the scarcity of fungicides registered for turf and, at the same time, in absence of alternative products to chemical ones. Moreover, unfortunately, for many years, turf management, especially in golf courses, has been based on excessive use of fungicides, without the support of experienced technicians and in lack of a correct diagnosis.

The greater availability of resources typical of golf courses, compared to traditional companies, has favoured for years an excessive use of chemical means. Training of technicians in this specific field has started recently (see boxes page 155 and 156).

© The Author(s), under exclusive license to Springer Nature Switzerland AG 2021   153
M. L. Gullino, *Spores*,
https://doi.org/10.1007/978-3-030-69995-6_34

Today sustainable management of turfgrass diseases is more common than in the past (see box page 157), but the use of more environmentally friendly methods on a large scale requires the widespread presence on the territory of well trained and constantly up-to-date technicians. And, above all, we need a greater and more widespread sensitivity: have you ever wonder why presidents of various football clubs do not hesitate to pay million-dollar salaries to the most famous football players and then skim off the few thousand dollars needed to properly manage stadium playground?

Curiosity

Some figures about golf

Golf is now a sport available around the globe, though geographically concentrated in the ten top golfing countries. In 2018, USA ranked first, with its 16,752 courses, followed by Japan (3,169), Canada (2,633), England (2,270) and Australia (1,616). Italy ranks 19th. In 2000 the courses were 267, with 59,000 players. In 2019, courses were 386 and players grew to more than 90,000. Although the sport has traditionally been associated with private clubs, where play is restricted to members, in fact worldwide golf is more and more played at public facilities.

The Italian turfgrass doctor

In 1992, after a stay in the United States at the Pennsylvania State University to specialize in turf-grass diseases, I sought a young graduate eager to work with me on the subject, at the time very neglected in our country. Massimo Mocioni, who had just completed his thesis work on a carnation disease, volunteered with great enthusiasm. He really had a nose for this topic.

Working with passion, Massimo Mocioni has become one of the best technicians in the sector, not only in Italy, and his deep knowledge of the secrets of sport fields and his approach to sustainable control strategies make him a reference point for many vips who turn to him for the care of their parks and private gardens.

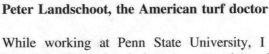

Peter Landschoot, the American turf doctor

While working at Penn State University, I was hosted by Peter Landschoot, a researcher with a double specialty in agronomy and plant pathology. Thanks to Peter I learned a lot about turfgrass, at a time when in Italy not so much attention was paid to it. The cooperation and friendship with Peter continued throughout the years.

Also turfs can be sustainable

In Italy turfs have a high turnover: we have in fact 40,000 hectares of sport and recreational green, 195,000 hectares of grassy road and railway areas, 250,000 hectares of agricultural areas and firebreaks, 100,000 hectares of public and private green areas. The whole corresponds to 2.5% of the national territory, with a management cost of 5–8 Euros per inhabitant.

Throughout a LIFE European project, involving several golf courses on the Ligurian coast, a few years ago guidelines have been developed for the adoption of environmental management systems. The project has focused on the sustainable management of golf greens, in particular as regards fertilization and disease management, the identification of biodiversity protection methods, the assessment of the biological quality of land and the possible use of renewable energy sources.

From Waste to Resource

Urban and industrial waste management is an ongoing major issue for environment and economy, involving not only public authorities but also enterprises.

A process that has long been used to recover waste is composting. Separate collection of organic waste and its treatment in composting plants has spread rapidly and today in many cities a large part of urban waste is transformed into compost of higher or lower quality (see box page 161).

Industrial waste of different origins, as well as green waste, need to be converted into compost. One of the main problems of compost is that it is often very variable in composition, depending on its origin, and this greatly compromises its possible usage in agricultural sectors (for example in horticulture) of some interest (see box page 162).

Many composts, in fact, while respecting the chemical and physical parameters required by law, are not usable because only partially mature. Compost, however, is not always usable as it is, because of the high concentration of soluble salts, the physical structure that rapidly changes during its use in the field, and the high pH.

The use of certain industrial waste, residues such as nut, hazelnut and bark processing, may contribute to alleviating some of these problems by improving drainage, or, conversely, by encouraging water retention of compost. It should be noted that, particularly in horticulture, the use of compost is acceptable only when mixed with peat or another substrate at a rate of 25–50%.

The use of compost as such appear to be possible as a fertiliser or soil improver, while when used as a substrate for horticultural crops it must be considered among the components and carefully dosed in relation to the specific crops it will be used for and the characteristics intended to be attributed to the end product.

Moreover, some composts have shown interesting suppressive ability against important pathogens of horticultural crops, highlighting a further exploitable aspect. In certain cases, physico-chemical characteristics and/or the presence of antagonistic microorganisms in some composts make them able to counteract diseases caused by several soil-borne pathogens.

M. L. Gullino, *Spores*,
https://doi.org/10.1007/978-3-030-69995-6_35

Compost production and use is in full development, being a resource of great interest for a modern multifunctional agriculture, able to enhance materials, such as organic waste, otherwise unusable (see box page 162).

In 2018, the total waste generated in the EU-27 by all economic activities and households amounted to 2,317 million tonnes. The agro-food sector contributes with waste and by-products with different characteristics depending on the specific supply chain from which they derive. Winemaking, for example, generates different by-products, including lees and pomace, than can be recovered for alternative uses than the production of alcohol in distilleries, such as for agronomic and energy purposes.

Waste comes also from fresh hazelnut processing—hard shells for example, characterized by a woody consistency, due to the high presence of lignin. Hazelnut shells, which account for 50% of the nut weight, can be used in the energy sector, for biofuel production.

The main solid waste generated in the rice production is rice husk, used as biomass or recycled for food industry or energy purposes. Among the sweet industry's wastes are cocoa processing residues, in the case of chocolate products, and wafer powder in the case of snacks.

Recent regulations allow nutrients recovery from anaerobic digestate of agri-food waste. Compost prepared with these residues has shown an interesting activity, not only for plant growth, but also for its pathogen suppression properties (see box page 163).

Going to depth

Waste production, a matter of rich countries

As countries develop from low-income to middle- and high-income levels, their waste management situations also evolve. Growth in prosperity and movement to urban areas are linked to increases in per capita generation of waste. Furthermore, rapid urbanization and population growth create larger population centers, making waste collection and the procuring of land for treatment and disposal more complicated. Urban waste management is expensive, representing often the highest budget item for many local administration in low-income countries, nearly 20% on average. In middle-income countries, solid waste management typically accounts for more than 10% of municipal budget, while it accounts for about 4% in high-income countries. Waste generation dynamically changes based on changes in economic development and population growth. With the economic crisis, also urban waste production falls: for instance, in Italy, between 2010 and 2011 it marked a -3.4% decrease, with a further decreasing of 7.7% in 2012. The world generates 2.01 billion tonnes of municipal solid waste annually, with at least 33% of it not managed in an environmentally safe manner. Worldwide, waste generated per person per day averages 0.74 kilogam, ranging from 0.11 to 4.54 kg. Though they only account for 16% of the world's population, high-income countries generate about 34% of the world's waste. Adequate waste disposal or treatment is almost exclusively the domain of high- and upper-middle-income countries.

Let's compost!

The success or failure of a compost depends very much on the type of waste used, on the level of maturation achieved and on the process used. Quality control of waste is crucial: a quality product must meet all waste acceptance criteria as verified by proper agronomic and physical–chemical analysis.

Like an ANT: hard work pays off

After years of research carried out as part of national and European projects by the plant pathology group of Torino, Massimo Pugliese, a young and enterprising researcher of Agroinnova, University of Torino, started a few years ago an academic spin-off. What is a spin-off? It is an enterprise born closely connected to the University, where it is "incubated" for some years, before becoming a real enterprise. A way of helping young people to realise entrepreneurial ideas and provide them with job opportunities. Massimo's enterprise is called ANT (that stands for AgriNewTech), referring to Massimo and his younger colleagues' industriousness. ANT produces highly effective compost-based fertilizers enriched with microorganisms that seem to work wonders on plants.

Stop to pathogens

In some cases, compost, in addition to agronomic interest, shows an interesting suppressive activity towards some soil-borne pathogens. This disease suppressive property would be due not only to antagonists' action, as seen for suppressive soils (see page 120), but also, in some cases, to the presence of particular chemical compounds that are produced during the composting process. This kind of compost is mainly used in the nursery sector.

The Good and the Bad of Pesticides

Consumers' negative attitude to pesticides is well-known. These chemical products used for plant disease management are often considered "poisonous" for human health. Yet they are not poisons, just agrochemicals (see box page 167), which, after many years of rigorous studies, as it happens with pharmaceuticals, are registered and made available to farmers as control means against pests threatening agricultural crops (see box page 168).

In many cases, the use of pesticides remains the most effective strategy, sometimes the only one, to reduce yield losses, both in quality and in quantity, caused by pests.

We are considering here fungicides, which are the pesticides used against fungal pathogens, trying to highlight the research needed for the development and use of new products. What we are discussing here, however, applies in principle to all pesticides, that is, also for insecticides and herbicides.

The probability of "discovering" a new pesticide through normal screening programmes has varied over time with the following progression: one successful active ingredient out of 5,000 substances tested in the 1950s; out of 10,000 in the 1960s; 20,000 in the 1970s; 40,000 in the 1980s; 50,000–70,000 in the 1990s; 100,000–120,000 in the 2000s. Currently, the commercial development of a plant protection product takes 8–10 years of research and costs, on average, 250–300 million Euros. This is because more and more chemical, biological, toxicological and environmental studies are needed to develop a new agrochemical, also by understanding its mode of action (see box page 169) as well as possible negative side effects, such as the development of resistance (see box page 170).

These data, together with considerations on the economic returns generated by individual products, mean that the agrochemical industry, while developing a new active ingredient, increasingly focuses its attention on the few major staple crops grown worldwide (such as cereals, apple, grapevine etc.), leaving aside the so-called "minor" ones (see page 173) which do not ensure a potential market of adequate size compared to the large investments required in research.

© The Author(s), under exclusive license to Springer Nature Switzerland AG 2021
M. L. Gullino, *Spores*,
https://doi.org/10.1007/978-3-030-69995-6_36

Together with the search of new active ingredients, an important role in crop protection has been played by the search of new formulations. In addition to traditional dry and wettable powders, concentrated emulsions and suspensions or other highly innovative formulations have appeared, which not only improve the distribution of agrochemicals, but also reduce the risk of phytotoxic effects and toxicity to humans and animals, and advance ease and safety in preparation, packaging, transport and handling.

Over time, the very concept of crop protection has undergone a profound evolution, which is also reflected in the terminology adopted. It is sufficient, in fact, to consult the texts of a few tens of years ago to witness a very frequent use of the term "pest *control*", in the sense of fighting, subsequently more often replaced by "crop *protection*" and then "disease *management*". This shows a real change in the approach, from pest eradication to their management, so as to avoid exceeding the tolerable thresholds for each crop and crop context.

Crop protection has therefore evolved from the prevalent, if not exclusive, use of control means (in the past most chemicals) to an integrated approach for disease management, based on the combined and rational use of the different means available to maximise the benefits and minimise the risks involved.

Pest control itself has gone from exclusively prevention strategies with "calendarized" chemical treatments (which in the past were the only possible due also to the unavailability of systemic products) to more targeted treatments, to be carried out on the basis of the actual occurrence of the disease and/or the careful assessment of the damage risk. The success of these latter measures is primary linked to the possibility of predicting infections and to the availability and effectiveness of registered fungicides able to act against ongoing infections. It should be noted that this evolution, which is still in progress, has been achieved also thanks to the remarkable evolution of cultivation techniques, increasingly sophisticated and aimed at reducing damage from pest attacks, environmental impact and production costs.

Integrated strategies for disease management have to be planned taking into account every crop context. Consider for example the different choices to be made in different soil and climate environments, in traditional cultivation systems or in organic farming, or in industrialized countries compared to developing ones.

A contribution to rationalizing use of pesticides has been provided by EEC Regulation n. 2078/92 of 30 June 1992, whose A1 measure—"Significant reduction in the use of plant protection products"—provided economic incentives for companies which undertake to comply with a product specification defined by a national committee on the basis of the specific environmental and cultural characteristics of each region. More recently, European legislation on the sustainable use of pesticides has entered into force, to further safeguard consumers (see box page 171).

The topic remains controversial. Pesticides, always subjected to years of careful study before being marketed, are still indispensable in many situations. At the same time, it must be clear to the consumer that plant disease management is more and more directed towards an increasingly limited use of pesticides, which are more and more sophisticated in their action mechanisms. Finally we must not forget that

European agri-food products, thanks to the widespread use of integrated disease management strategies and the several controls, are among the safest in the world, as clearly shown by the data regularly made available by the European Food Safety Agency (EFSA) (see box page 172).

Pesticides are not poisons!

If we all get used to considering pesticides as real drugs for plants, indispensable in some circumstances, right as medicines are essential for human health, we would avoid making pointless terrorism by calling them "poisons", also because, for consumer protection, food-safety regulation has imposed strict control procedures on the possible residues of chemical products in foodstuffs.

What are pesticides? How are they developed?

Pesticides, also called agrochemicals, are active substances used to control pests (insects, plant pathogens, fungi or weeds).

The search of new active ingredients today complies with different criteria than those of the past: in addition to random screening, increasingly we rely on the so called "imitation" chemistry, which consists in the synthesis of products similar to others already in existence, the evaluation of natural products and the obtaining of their derivatives up to a biorational approach.

Imitation chemistry is based on the synthesis of new molecules, very similar to others already patented, but characterized by higher biological activity and/or a wider activity spectrum. Sometimes, the starting point is made up of substances of natural origin—which serve as a model for the development of more or less markedly modified analogues. The biorational approach, widely adopted in the pharmaceutical sector, is based on the development of substances capable of inhibiting key enzymes or metabolic processes indispensable for the pathogen.

Since the second half of the 1980s there has been a substantial reduction in the synthesis of new molecules, as a consequence of the severe restrictions on their registration.

Some history of fungicides

Going to depth

The history of fungicides is made up of three phases, characterized by the finding and development of new molecules. The first one, called sulphur era, goes from ancient times to 1878; the second one, the copper era, from 1878 to 1934; the last is the synthetic fungicides era, from 1934 to the present days. Since the 1940s, important advances in knowledge of plant physiology and biochemistry led to studies that initiated the sub-era of endotherapic fungicides in the late 1960s. These fungicides, moving inside the plant vessels, have several advantages: resistance to water washout, ability to interfere with ongoing infection processes (potential curative action), distribution in the various parts of the plant based on the intensity of the leaf transpiration, high biological activity, effectiveness at low dosages, selectivity. Moreover, they are less toxic to human and animal health and have a low environmental impact. One of the prerequisites for endotherapic activity is a high selectivity of the toxic action, which must be carried out at the expense of the target fungus and not towards the host cells. Such selectivity is allowed by a very specific action mechanism at the cellular level, which occur at individual metabolic sites (monosite fungicides), unlike the traditional protective fungicides that have usually a nonspecific mode of action. The modern evolution of fungicides, marked by an increasing attention to environmental impact and to human and animal health, has thus been oriented towards the use of molecules with specific action mechanisms, effective already at the lowest doses and less dangerous for humans and environment.

Fungicide resistance

The multiple benefits afforded by a specific mode of action are offset, as a main disadvantage, by the risk that targeted microorganisms become resistant to that particular fungicide. As has happened in human and veterinary medicine in the case of bacteria, which with great ease became resistant to antibiotics, at the end of the 1960s the appearance of some strains of certain fungal pathogens resistant to specific fungicides exploded in all its severity. Why did this phenomenon occur at the end of the 1960s and not before? After all, fungicides had been introduced a long time before! For a very simple reason: marketing and repeated use of monosite fungicides, able to interact with mechanisms very specific to that fungal cell has favoured, inside the population of certain pathogens, the selection of some resistant strains, capable of developing and causing damage to plants even in the presence of treatments proved to be effective until shortly before. In other words, resistant strains are mutants, characterized by the presence of mutations at the target-site where the fungicide acts. Statistically, it is more likely to develop resistance towards monosite fungicides, as it requires a single mutation, rather than towards fungicides able to act on various sites of the fungal cell. In fact, the appearance of resistance has affected different classes of fungicides and different pathogens, creating, especially in the years 1960–1970, not a few practical problems. Several chemical products used on economically important crops became no more effective, especially in the cases where a large number of treatments was carried out.

Fungicide resistance is among the most challenging issues of modern plant disease management, due to its important consequences on all stakeholders involved in horticultural crop production: agrochemical industry, for the loss of effectiveness of molecules requiring ever greater investments in research and development; farmers, for yield losses caused by diseases no longer controlled by pesticides; and consumers, for the price increase caused by higher production costs and reductions in supply.

How to protect consumers

The increasing demand of quality assurance and food safety systems by international markets and large-scale food retail, for some time have strongly influenced protection strategies, especially for high-income crops and export products.

Both commercial operators and farms are now increasingly called upon to adopt HACCP (Hazard Analysis and Critical Control Point) and ISO (International Organization for Standardization) standards, voluntary certifications of product or company management system (quality, environmental impact, food-safety, etc.), and to comply with product specifications and protection of origin or trademarks. These two certification forms (product and system) are intended to evolve, integrating into forms of "system and product" certification, which are intended to represent the most complete form of quality assurance both in the cogent and voluntary sector.

This is proved by the growing demanded evidence of the voluntary EUREPGAP (European Good Agriculture Practices) certification, as a prerequisite for doing business. This certification covers exhaustively and even too strictly the identification of critical and control phases in the whole production process, from planting to harvest and marketing of horticultural crops, providing an assurance on the farm management practice, which concerns products'quality, but also environment protection and work quality of the operators themselves.

Overall, the adoption of the various quality assurance schemes can only be viewed positively, as it leads the agricultural sector to adopt eco-friendly and transparent production processes and to provide the consumer with quality guarantees, food safety and product traceability.

Residues? No, thanks!

Another negative aspect deriving from the use of endotherapic fungicides is the possible presence of residues in foodstuff, a phenomenon particularly important in the case of leafy vegetable crops (lettuce, for example). However, the correct use of pesticides, at the dosages indicated on the label and in compliance with the interval that must elapse between the last treatment and the harvest, allows to avoid an amount of residues higher than the maximum concentration legally permitted.

Moreover, all the analyses carried out show that the vast majority of agricultural products contain pesticide residues far lower than the maximum residue limit. In Europe, the European Food Safety Agency (EFSA) data, show that produce are among the safest: less than 2% of samples are irregular, containing pesticide residues higher than those legally admitted, while more than 40% of the tested products does not even contain any residues detectable with the current available methods.

Minor Crops of Great Importance

At global level, crops are divided into "main" and "minor" ones: the former (rice, cotton, wheat, soya etc.), a dozen in all, are those that interest very large acreages, while the latter are less globally widespread.

Minor crops require, at a global level, a negligible use of technical means for pest control compared to what it is necessary for the major ones.

This means that, especially at a time when pesticide development and registration has become increasingly complex and costly, the agrochemical industry worldwide has very little interest in keeping their products registered for use on minor crops or to register new minor use pesticides.

In addition, the process of re-evaluation of old pesticides has led to the loss of many of them, often because no longer economically attractive, due to the presence of newer products on the market. That is why minor crops are more and more "pesticide-orphans" and, consequently, "prone" to pest-attacks, and growers are helpless. Moreover, if in the recent past the pesticide companies found insufficient economic incentive for developing products on these crops, they are even less interested in doing it now, with increasingly stringent registration rules.

Where is the problem—you might wonder. Unfortunately, or better luckily, Italy as well as other Mediterranean countries, is characterized by the presence of a wide variety of genera and cultivated species! Our diet, in fact, is not made of a few staple crops, such as potato, cabbage, turnip and little more, as happens in some countries of Northern Europe. Our tables are often filled with peppers, aubergines, chicory, rocket, basil, apricots, plums, almonds, prickly pears, mint, small fruits etc. It is estimated that we consume regularly at least 160 cultivated species.

Therefore, for many Mediterranean countries, Italy included, minor crops are of considerable economic importance and, unfortunately, they are subject to several pest attacks. Disease management in minor crops is therefore a very serious problem for Mediterranean agriculture. At Community level, north-European countries have relatively more tools for protecting their crops: the monotony of their production, in fact, make them insensitive to minor crop issues.

M. L. Gullino, *Spores*,
https://doi.org/10.1007/978-3-030-69995-6_37

One possible solution is carrying out, in the most critical cases, the needed experiments to produce the documentation required for obtaining an extension of use of pesticides already registered on similar crops (see box page here below). The lack of registered active ingredients remains a serious issue for many of the minor crops—yet of great importance—of Mediterranean production.

Curiosity

Mobilization for basil

More than once, as a result of the emergence of new pests affecting basil crops, In Italy, growers and local authorities have taken action, by supporting Research Centres in order to carry on the needed trials aimed at obtaining the label extension for products already registered on similar crops such as lettuce. In these cases, Research centres carry out the trials needed to assess pesticide effectiveness and the presence of any residues, then fill the documentation required to get the label extension from the competent Ministries. This approach, which is considered by the current legislation, permit agrochemical companies to extend the registration of their products at very low cost and growers to have a number of chemicals available in the most critical situations.

A Small Contribution to Fill the Ozone Hole

Once upon a time there was methyl bromide... If it were a fairy tale it would start like this. But it is not a fairy tale. Or maybe it is. Because this topic has really involved twenty years of my life and, a bit like it had been, in the eighties, fifteen years before with grey mould, it left its mark.

It is the story of methyl bromide, a fumigant widely used in the past to control soil-borne pests in intensive horticultural systems, often carried out in greenhouse, in absence of or with limited crop rotation.

Methyl bromide is a very toxic product not only for plant pests but also for humans and, for this reason, it was applied only by specialized and certified personnel (the so-called fumigators). In the late 1980s, researchers showed that methyl bromide, as well as several substances used in refrigeration systems (the notorious chlorofluorocarbons), throughout the products of its degradation in the atmosphere, contributes to stratospheric ozone depletion.

Stratospheric ozone protects us from the negative effects of ultraviolet radiation; some countries, more than others, close to the poles, suffer very seriously the ozone hole consequences. So, in the late 1980s, it began the countdown for this fumigant, and an exciting professional adventure for me. As all the global issues, also the phasing out process of the production and use of substances responsible for ozone depletion—including methyl bromide—has been handled by the United Nations as a multilateral agreement, then ratified by the Montreal Protocol (see box page 177). It was back in 1992 that my research group was entrusted, by the Italian Ministry for Environment, with the task of following this issue for our country, because the reduction in the consumption of methyl bromide was concerning our agriculture. I threw myself headlong in this real adventure, understanding how the results of the research carried out for years by plant pathologists at the University of Torino could be useful to legislators. Attending and actively participating in the negotiations amused me a lot. The various countries involved had quite different opinions on the topic (see box page 178). Excited by this new experience, I felt a bit like a homeland savior.

Until 1991, Italy was the world's second largest consumer of methyl bromide as a very strong producer of horticultural crops. Our country had to make every effort to comply with the scheduled phasing out in the use of this fumigant as imposed by the Montreal Protocol and, thanks to the coordinated action of growers, extension services, researchers, agrochemical industry, fumigators and public administration, succeeded within a few years, reaching the total elimination (phase-out) of methyl bromide in 2005. In the 1990s and early 2000s there was considerable international and national commitment to search and develop alternatives to methyl bromide for soil, substrate and foodstuff disinfestation, effective under different situations. Thanks to the efforts of all, Italy—once the tail light at world level for its huge use of methyl bromide—managed to drastically reduce the usage of this fumigant, thus establishing itself as a reference point for many countries still struggling with the phasing out. In quite a few years, Italy has reversed its position, reaching another record, that of having made available on the national market virtually all the known alternatives to methyl bromide. Thanks to many national and international projects, non-chemical alternatives have been adapted to our growing conditions, while efforts have been made to rationalise as much as possible the use of pesticides and minimise their environmental impact (see box page 178).

In the face of a global environmental emergency, Italy has therefore maintained a serious commitment to eliminate a fumigant that was really contributing to stratospheric ozone layer depletion.

After having addressed its agricultural issues, providing farmers with diverse solutions for the various contexts, Italy has strongly committed itself in many countries (China, Morocco, Romania, Kenya, etc.) where it has intensively cooperated in the transfer of the results obtained, so to help them to meet the 2015 deadline for the methyl bromide phasing out for developing countries.

International agreements on global issues

All issues that have a global impact (climate change and ozone depletion, for example) must be tackled in such a way as to take into account the needs of every country, but with the final aim of taking timely decisions for the benefit of all. In these cases, a supranational organisation, such as the United Nations, must take action in order to accelerate the environmental protection processes by mediating the various needs.

Generally, the needs of developing countries or emerging economies are particularly protected under the various Protocols. For these countries, globally adopted restrictions provide for longer implementation times (the so called grace period), to allow them to meet the goals through appropriate funding and with the help of the industrialized countries.

The Montreal Protocol is focused on the reduction in the use of substances harmful to stratospheric ozone, while the Kyoto Protocol deals with climate change issues.

Watch out for the best and the brightest

On the elimination of the use of methyl bromide, at international level, especially in the early years, Italy has often struggled in countering the positions of some north-European countries ready to give up the use of the fumigant. In fact, some countries had already phase out this fumigant much before the discovery of its negative action on the ozone layer, because it polluted their very superficial groundwater. It was therefore easy for that country to eliminate the use of a product that was no longer used. What however remains unknown to consumers is the fact that, at the same time, some of those countries moved their tomato production in northern Africa, where they kept on using methyl bromide, selling the tomatoes produced there on the European market.

Going to depth

Europe requires more sustainable fumigants

Several European projects have addressed the controversial issue of protecting intensive crops from soil-borne pathogens and pests without the use of methyl bromide, taking into account that the European regulation on the registration of new pesticides is particularly stringent and does not allow for a rosy future for chemicals in general and for fumigants in particular.

Over the years, there has been a gradual shift from exclusive use of fumigants to their use combined with non chemical means, using an integrated approach.

Good Guys Against Bad Ones

The first research on the use of biological control of plant pathogens dates back to the 1950s. Biological control is based on the use of microorganisms (or their products, such as enzymes) against pathogens—a sort of fight between good guys (antagonists) and bad ones (plant pathogens). The search for biocontrol agents, started slowly and then became increasingly widespread, involving different fields and sectors and has been strongly encouraged in recent years by the increasing difficulties encountered in the registration and use of traditional pesticides. Finding effective biological control agents is far from easy, since the interactions between antagonistic microorganisms and plant pathogens are quite complex. Moreover, in some cultivation systems highly conditioned by environmental factors and human interventions, it is not always trivial to introduce a microorganism that, in order to exert its antagonistic action, has to colonize the environment (soil, phyllosphere, carposphere…). Let us think, for example, about the effects of rain and human interventions (other treatments, irrigation, etc.) on an orchard. But even in a greenhouse, things do not get much better.

Indeed, the poor results often obtained in the development and practical use of biocontrol agents are explained by the extreme complexity of the systems in which these microorganisms have to be introduced.

However, the scarcity of chemical resources and the increasing difficulties in registering them require a continuous search for effective biological means. So, what are those already developed, and how do they act? Among the biocontrol agents developed so far there are bacteria and fungi that, in the course of several researches, have proven to be effective by controlling important pest attacks.

Among the bacteria, the strain K 84 di *Agrobacterium radiobacter*, stands out for its effectiveness against *Agrobacterium tumefaciens*, causal agent of bacterial tumors on various species (grapevine, rose, etc.) thanks to its ability to colonize at root level the sites that would be colonized by the pathogen and also to produce bacteriocines, antibiotic-acting substances that inhibit the development of the phytopathogenic bacterium.

M. L. Gullino, *Spores*,
https://doi.org/10.1007/978-3-030-69995-6_39

179

It is almost a war between equals, between two microorganisms systematically very close, which for this reason occupy very similar ecological niches. The secret to success, in this case, is that the antagonist reaches the right place before the pathogen. Another example of success in biological control is represented by fluorescent strains of *Pseudomonas*, often isolated from the soil, able to counteract several pathogenic fungi responsible of cereal foot-rot diseases and various other rots. Also the selected fluorescent *Pseudomonas* strains act with a double mechanism: by colonizing the sites that would be occupied by the pathogen and by producing antibiotic-acting substances, capable of inhibiting the development of the different target pathogens.

The study since the 1950s of some species of *Trichoderma* antagonistic fungi has led to the development of registered strains used as biocontrol agents in some countries. Some strains are effective against causal agents of foliar diseases (such as the already mentioned grey mould, see page 71), while other strains are able to counteract soil-borne fungal pathogens (*Rhizoctonia solani*, *Sclerotinia sclerotiorum*, etc.). *Trichoderma* acts against different pathogens with different mechanisms. In the case of grey mould, for example, the antagonist competes with the plant pathogen (*Botrytis cinerea*) by colonizing the attacked sites in advance, using the same nutrients that the pathogen would use. Against *Rhizoctonia solani*, however, the antagonist acts by becoming a parasite of the pathogen itself, even using the pathogenic fungus as nourishment: this phenomenon is called hyperparasitism. You see, *Trichoderma* is a very busy antagonist! There are several commercial formulations prepared with various *Trichoderma* strains, used on different crops (grapevine, tomato, ornamental, etc.) and many researchers around the world working on such an antagonist! (see box page 181).

Biological control means find interesting possible post-harvest applications, against fungi responsible of that fruit rot we all know. Yeasts are among the most effective means against the development of fungal species belonging to *Penicillium*, *Botrytis* and *Alternaria*, that attack apple, pear and citrus crops. These yeasts act mainly by competing with pathogens in the colonization of micro-wounds present on the fruits.

It is expected that the use of biocontrol agents will increase in the coming years, also thanks to the continuous reduction of chemical means available.

The *Trichoderma* guys

Since the seventies a kind of "*Trichoderma* club" started, made up of researchers working in the development of *Trichoderma* strains, able to counteract the development of plant pathogens. This bunch of researchers met every 4 years in order to exchange data, information and build effective partnerships. Among them, Gary Harman, at Cornell University, USA and Matteo Lorito, from University of Naples, Italy. The three of us even spent a period of research all together at the Geneva Campus of the Cornell University, in the late 1980s.

From Solar Cells to Tomato

The last 30 years have been marked by the search for control means alternative to synthetic chemical ones. The attention of researchers has also focused on the possible use of so-called "natural" products, such as, for example, some inorganic salts, interesting for their ability to reduce the development of various plant pathogens. Some of these, such as sodium bicarbonate, are already widely used in the food industry and have good characteristics for inclusion in integrated disease management programmes: low cost, favourable safety profile for humans and environment, low toxicity to mammals.

Also the use of silicates is very interesting: they act against some pathogens both stimulating natural defense mechanisms in the plant, strengthening the cell walls, thus creating a natural barrier that prevents pathogen penetration into the host plant tissues.

In the past, silicates were tested mainly for protection against attacks of fungal pathogens on cereals, rice in particular. In recent years, experiments have been made with silicates also in the case of vegetable crops, grown in soilless cultivation systems (see page 121), with very interesting results. The treatment with potassium silicate allows, in fact, to control the development of different pathogens, responsible of foliar (powdery mildew, downy mildew etc.) as well as vascular diseases.

A promising evolution for plant disease management is the use of silicates deriving from industrial processes. In fact, silicates are used in various industrial processes, such as the production of solar panels and electronic devices, generating several by-products whose recovering is useful and profitable.

© The Author(s), under exclusive license to Springer Nature Switzerland AG 2021
M. L. Gullino, *Spores*,
https://doi.org/10.1007/978-3-030-69995-6_40

Trials carried out, in collaboration with industries in the sector (see box here below), have allowed to prove the effectiveness of silicate-based wastes towards various foliar diseases affecting tomato, zucchini and other vegetable crops. The recovery of silicates from the photovoltaic industry can therefore be useful for pest control in both traditional and organic farming.

Going to depth

Silicates

Silicates are very interesting for their effectiveness against a number of plant pathogens, since they act as resistance inducers. Applied in the soil or substrates, as well as in soilless systems, at the nursery level, they provide a long-term effect and can be used on many crops, such as tomato, lettuce, pepper, etc.

Also Seeds Go to the Spa

Many pathogens survive from one cultivation cycle to another by means of seeds, which carry them internally or externally, and for this reason they are called seed-borne pathogens. Seed provides therefore an efficient means in disseminating plant pathogens in vegetable crops. It often takes just one infected seed out of 10,000 used at the time of sowing to have serious damage in the field. We have already seen many examples of pathogens transmitted and spread by infected seeds. This phenomenon, increasingly serious and frequent in our global trade era, affects minor (basil, rocket) as well as major crops (cereals, for example).

The use of healthy or treated seeds, in order to eradicate pathogens, with the various means available, represents a pillar of integrated disease management strategies. A great contribution can be provided by molecular diagnostics that can facilitate the health status of seed lots.

Both for the lack of synthetic chemicals and for the need to use methods with low environmental impact, the use of physical or biological means for treating seeds is increasingly widespread.

For years, seed treatments in hot water have been effective for bacterial and fungal diseases control. This kind of treatments, which have been neglected in recent decades to the advantage of fungicides in the case of fungal diseases, has found again interest because of the need of using healthy seeds in organic farming, where it is not possible to use chemical treatments. In particular, thermotherapy treatments with hot water provide interesting results to reduce bacteria contamination (see box page 186).

As a part of a project funded by the European Commission, an innovative machine has been developed in Sweden that allows the use of aerated steam. The treatment is carried out with an equipment consisting essentially of a first chamber in which the seeds are preheated, a second chamber in which the treatment is carried out and a third one in which the seeds are quickly cooled down.

Naturally produced substances from plants or microorganisms represent a huge pool of molecules with potential biocide activity. The search for substances of natural origin in seed treatments is very intense: oils and other extracts are able to

M. L. Gullino, *Spores*,
https://doi.org/10.1007/978-3-030-69995-6_41

provide partial protection against seed-borne pathogens of different species. Particularly interesting for their results on some pathogens are seed dressings with thyme or savoury oil at 0.1%, tested on certain horticultural crops: in the future it might be possible talking about real "seed massages". It is like a real spa treatment.

Some antagonistic microorganisms can also be used in seed treatments, yet not always with exciting results.

Fungicide seed dressings would allow to limit the spread of the main fungal pathogens in the world, however there are few chemical products registered for such use and restrictions imposed by the European regulation make the use of fungicides more and more difficult.

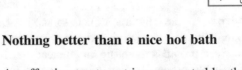

Nothing better than a nice hot bath

An effective treatment is represented by the immersion of seeds in hot water at temperatures varying between 45 and 50 °C for 20–60 min. Hot air treatments can also be effective: in the case of lettuce seeds, exposure to 70 °C for 1–4 days proved to be effective against certain phytopathogenic bacteria.

The Patient Work of Plant Breeders

The history of crop genetic improvement is full of results: think of the yield increments obtained in cereal crops thanks to improvement for bunt and rust resistance. Genetic resistance is the safest and most economical method for plant disease management, always and especially advisable in all cases where other means are not available or not suitable for technical, economic or environmental reasons.

In the case of low-income field crops (cereals, for example) or pathogen particularly difficult to control (i.e. viruses and soil-borne pathogens), the use of resistance is the most cost-effective solution. In the United States of America, 95% of cereal cultivars are resistant to at least one pathogen.

The fair levels of rusticity and pest resistance shown by some old cultivars, moreover, testify the long and often unconscious work of selection practiced in the past, even without knowing the resistance phenomenon nor, in some cases, plant diseases. The rational application of the selection within commonly cultivated varieties led, at the beginning of the 20th century, to obtain lines of various agricultural species endowed with a good resistance level against economically important diseases.

The discovery of the genetic nature and the Mendelian inheritance of resistance has given a new impulse to resistance use against dangerous pathogens in the most important cultivars by hybridization with other cultivated varieties or with similar wild species naturally resistant.

Pest resistance develops in nature where host and parasite interact for a long time exerting a mutual selection pressure, until establishing forms of coexistence (see box page 189).

The use of varieties characterized by monogenic or oligogenic resistance leads to very good results in field. However, experience has shown that often the results of years of work for the selection of resistance to diseases are quickly nullified by the appearance of new physiological races of the pathogen able to infect cultivars equipped with new resistance genes. This is what happened, for example, in the case of potato late blight.

M. L. Gullino, *Spores*,
https://doi.org/10.1007/978-3-030-69995-6_42

The countless cases of easy and rapid resistance overcoming have induced genetic breeders to focus more on the polygenic resistance. Experience has shown that this type of resistance, although less high, tends to have longer efficiency as it is more difficult to overcome. In the case of polygenic resistance, there are no differential reactions between host and pathogen genotypes, the level of effectiveness is generally low and linked to the environmental conditions and disease pressure. However, in the case of polygenic resistance, resistance manifests itself towards all races of a pathogen and is stable. The polygenic resistance, therefore, does not provide total pest control, especially under conditions particularly favourable to disease development, but it is well suited to be inserted in a context of integrated pest management.

To conventional breeding systems, based on crossings, recurrent selection and progeny selection can be added alternative systems, based on co-cultivation of plant and pathogen under *in vitro* conditions and on the use of pathogen toxins.

Genetic improvement has mainly concerned open field crops for which chemical control methods, even when present, are not always economically viable. Very intense has been also the selection of resistant cultivars for vegetables. In the case of fruit and forest species, obtaining resistant cultivars is much slower and more difficult (see box page 190). Finally, more tricky because of the very high number of cultivated species and the rapid changes of consumer preferences, is the search for resistant cultivars in the ornamental sector, where resistance has concerned only the most common species (such as carnation and bulb crops), and the pathogens most difficult to control, due to the lack of pesticides or to the low effectiveness of the available means. Very active, in this regard, has been the search for carnation and bulb crops cultivars resistant to *Fusarium* wilt. Ligurian plant breeders have proven to be among the best in this task (see page 89).

The use of resistance (especially horizontal) is very interesting in developing countries.

Grafting on resistant rootstock, though expensive, is interesting in the case of vegetables (see box page 190). Rotating crop cultivars with different resistance genes or using cultivars with various resistance genes for different areas can maximize vertical resistance effectiveness. Vertical resistance is applicable in the most advanced agricultural systems—where it is possible to resort to a rapid succession of cultivars equipped with different resistance genes—and in all those agricultural systems that can tolerate only limited damage. In Europe unconventional breeding throughout genetic engineering is not practiced, despite the new approaches now adopted (see box page 191).

Going to depth

Looking for resistance genes?

Where do geneticists go to look for resistant plant material? In the centers of origin of crop plants (which are also the same geographical areas where pests come from). From a practical point of view, the search for resistance sources should be made in priority order: among the local populations of the cultivar; among old indigenous varieties; among the wild forms of the same species; in other similar species; in other systematically related genera.

Let's see some examples: resistance sources of potato to *Phytophthora infestans* have been found in wild varieties of Central America (Guatemala and Mexico); for wheat, resistance to rust has been found in *Triticum timopheevii* grown in Middle East. Species of genus *Vitis* resistant to downy mildew come from North America, region where—as already mentioned —powdery mildew, downy mildew and grape phillox-era originated (see page 23). In the case of tomato, *Phytophthora infestans* resistance genes have been found in *Lycopersicon pimpinellifolium*, *L. hirsutum* and *L. peruvianum* species, grown in the Andean regions. In plant breeding for pest and disease resistance, one should always keep in mind that this goal cannot be disjointed from other crop improvement aims. Indeed, even a very resistant variety would be doomed to commercial failure if it did not have, at the same time, other desirable agronomic traits.

The trick to breeding is patience

The search for apple cultivars resistant to apple scab disease (*Venturia inaequalis*) has been particularly long and expensive: over forty years of research have led to the selection of a few resistant cultivars, including "Fiorina" and "Liberty". However, the selection of new races of the pathogen capable of overcoming the resistance succeeded rather quickly.

Going to depth

Resistant rootstocks

The use of resistant rootstocks allows to control several soil-borne pathogens causing root and vascular diseases, as an excellent alternative to soil disinfestation. They have been adopted for a long time in the case of fruit crops, for example in grapevine against *Phylloxera* and, in recent years, more and more frequently in the case of root and vascular diseases of vegetable crops—mainly for tomato, eggplant and melon.

Grafting has very high costs, higher than those of soil disinfestation, because it requires highly specialized labour. Beside these economic obstacles, sometimes there are also technical issues. For example, consumers tend to prefer melon varieties that are susceptible to Fusarium wilt rather than those grafted on a resistant rootstock. In some cases, moreover, grafting onto resistant rootstock causes the emergence of new phytopathological problems: European grapevine grafted on American rootstocks, for example, is susceptible to chlorosis, while citrus grafted on bitter orange resistant to *Phytophthora* are extremely susceptible to Citrus tristeza virus. As you can see, as soon as you solve a problem, here is a new one.

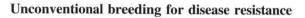

Going to depth

Unconventional breeding for disease resistance

Conventional breeding plays an essential role in crop improvement but usually entails growing and examining large populations of crops over multiple generations, a lengthy and labor-intensive process. Genetic engineering, which refers to the direct alteration of an organism's genetic material using biotechnology, has several advantages compared with conventional breeding. First, it enables the introduction, removal, modification, or fine-tuning of specific genes of interest with minimal undesired changes to the rest of the crop genome. As a result, crops exhibiting desired agronomic traits can be obtained in fewer generations compared with conventional breeding. Second, genetic engineering allows for interchange of genetic material across species. Thus, the raw genetic materials that can be exploited for this process is not restricted to the genes available within the species. Third, plant transformation during genetic engineering allows the introduction of new genes into vegetatively propagated crops such as banana, cassava and potato. These features make genetic engineering a powerful tool for enhancing resistance against plant pathogens.

Most cases of plant genetic engineering rely on conventional transgenic approaches or the more recent genome-editing technologies. In conventional transgenic methods, genes that encode desired agronomic traits are inserted into the genome at random locations through plant transformation. These methods typically result in varieties containing foreign DNA. In contrast, genome editing allows changes to the endogenous plant DNA, such as deletions, insertions, and replacements of DNA of various lengths at designated targets. Depending on the type of edits introduced, the product may or may not contain foreign DNA. In some areas of the world, including the United States, Argentina and Brazil, genome-edited plants that do not contain foreign DNA are not subject to the additional regulatory measures applied to transgenic plants. Regardless of differences in regulatory policies, both conventional transgenic techniques and genome editing continue to be powerful tools for crop improvement. Needless to say, transgenic plants are not very popular in Europe.

Diagnosis? Fast, but Not Too Much

In our profession as plant pathologists, it can really be said that everything begins with diagnosis. Indeed, to put it better, everything begins with a *good* diagnosis. You can apply to plants what is valid for humans and animals. Both a good doctor and a good veterinarian, before treating their patient, have to understand and accurately identify the cause of the disease. The same applies to plant pathologists, who compared to the medical doctors and the veterinarians, have some advantages and some disadvantages. The main disadvantage is that plants do not speak and, therefore, cannot help us to understand and interpret their symptoms. The main advantage is that, even in the extreme case of our inability to understand and solve the problem, if the "patient" ever dies, the damage is never so dramatic as in the case of animals and humans.

A timely diagnosis is therefore one of the most important step in plant pathology. Diagnosis of plant diseases requires, in the simplest cases, the use of quite simple instruments (magnifying glasses, microscope...), yet more frequently it requires laboratory steps, with more or less complex techniques.

Before the diagnosis, actually, there is sampling in field or greenhouse of the infected plants or part of them that will be analysed. This is a fundamental and very critical step (see box page 194). Collected samples should be kept in paper bags rather than plastic, to prevent them from rotting, and sent very quickly to the lab you want to consult.

You cannot imagine how many unsuitable samples, absolutely insignificant, if not rotten, we receive... Only in the case of relatively common pathogens and greatly experienced lab technicians, it is possible to identify a pathogen by simple visual examination (see box page 195). Generally, and especially in the case of fungi, microscopic observations and, very often, pathogen isolation from infected tissues with artificial inoculation on suitable substrates in Petri plates are required. More complex to identify are, in increasing order, bacteria, viruses and phytoplasmas. In the case of the two latter ones, observations with electronic microscope or very advanced diagnostic techniques based on molecular methods are required (see box page 196).

Often, especially in the case of completely new pathogens, never previously reported, it is necessary to reproduce the symptoms on healthy plants. I mean, without going into too much detail, it is clear that identifying the agent of a disease is complicated and time-consuming. Also because, as already mentioned, plants do not speak and therefore cannot provide us with useful information. Also because, often, different pathogens cause similar symptoms. Certainly the more experience you gather, more likely you will formulate the correct diagnosis. Today, new technologies allow us to make diagnosis faster, using serological or molecular techniques very similar to those used in medical diagnostic laboratories. It is important to underline that these techniques, very useful and often decisive, give the best when they complement direct plant observations in the field or on collected plant samples. Recently, very useful portable instruments have been developed, in order to permit to carry-on the diagnosis directly in the field.

Having to do with a plant, in fact, not being able to collect useful information from the "patient", infected plants observation, possibly in the field, remains of great importance. Nowadays, the new technologies also permit long-distance diagnosis, made just looking at photographs… (see box page 196).

Going to depth

Sampling: a very critical step

Samples collection for further analysis by the expert is a crucial stage significantly impacting on diagnostic success. The sample (the whole plant, one of its branches, some leaves, fruits), must in fact be "representative" of the situation observed in the field. Plant or parts thereof with symptoms should be collected at different stages to allow the technician to observe the evolution of symptoms and to choose the right material to be used for observations and analysis. Moreover, in collecting the sample to be sent to the expert, it is appropriate to make some observations on the localization of symptoms in the field, on how the infection expands, etc.

Nose knows

There is no Ligurian grower who does not know Stefano Rapetti. He is not an expert in perfumes and not even an oenologist, even if, thanks to his origins from Acqui Terme, where the Italian DOCG Brachetto d'Acqui wine is produced, he does not despise good wine. Yet Stefano's nose is directed rather towards plant diseases, especially those of flowering crops. Stefano Rapetti has assisted for years the Ligurian floriculturists (actually he keeps on doing it full-time even after his retirement), maturing an experience more unique than rare. There is no flowering crop or cultivar—along with all its pests—that Stefano has not known personally. Symptoms descriptions he is able to offer by voice or by writings sound like poetry to the ears of a plant pathologist: nobody can find the right term, the most suitable adjective to describe a symptom or a variation of colour better than him. His words exude competence, experience, passion.

Molecular methods

In the last 25 years, research in biotechnology has led to the development of molecular methods for the diagnosis of plant diseases caused by viruses, bacteria and fungi that rely on nucleic acid hybridisation and DNA amplification by polymerase-catalyzed chain reaction (PCR) and other even more recent methods. These techniques are very sensitive and specific and are more and more extensively adopted. Nowadays, some of these methods, combined with up-to-date equipment, can be used directly in the field.

Seeing first-hand: observation's priority before diagnosis

A famous American phytopathologist, J. C. Walker, in the first half of 1900, in addressing recommendations to younger colleagues, reminded them of the importance of going in the fields and observing with their own eyes the infected plants, their symptoms, instead of just staying in the lab. Today, in the internet era, it is fashionable to do everything remotely. Many of our younger researchers know only the DNA of pathogens and, as far as plants are concerned, they do not recognize an apple from a pear tree. Sitting all day at the computer, not only they do not go in the field but also in experimental greenhouses that are a few tens of meters away. What shall I say? If I were a sick plant, I would never want to end up in their hands... Dinosaur word, maybe.

Biosecurity: From Lemons of Grandma Olimpia to Present Research

Biosecurity is among my latest research topics. Who would have ever imagined it in the far late 1960s! At that time, in fact, I remember that I was impressed by Uncle Piero's sharp warning to grandma Olimpia, who was about to leave for United States to visit him—a historic undertaking at that time for an over seventy years old woman from the small town of Saluzzo, who did not speak a word of English! Well, Uncle Picro, knowing that his mother was very proud of her magnificent lemons that she had managed to grow in Saluzzo, recommended her to absolutely not carry them in the suitcase, because the strict customs controls in use since then at the American airports would have intercepted them, causing problems.

He explained to me that Americans did not want alien pests to be introduced into their country, providing—as does every good researcher—any possible detail that would satisfy my curiosity of thirteen years old.

Without knowing, I was then approaching today's much debated issue of biosecurity. By the way, grandma ignored that wise warning and still packed her lemons, which obviously ended up being intercepted by customs at the airport arrival! Later in the years, I learned how important it was and still is for different countries to prevent the entry of plant pests through passengers and goods. Every time I went to United States—which was very common to me since the 1980s, my American years—and was asked to fill the entrance forms to the country, I felt very guilty to deny having visited farms in the 15 days before the journey. However, if I had ever declared my daily attending not only of agricultural companies but, above all, of experimental greenhouses and infected plants, I would still be stuck at the first visited airport customs!

Today biosecurity is my daily bread, since globalization has led to a now unstoppable transit of goods from one country to another: today we consume fruit and vegetables produced in other continents and use seeds and planting material imported from other countries much more easily than in the past.

This globalization of trade comes with a corollary of effects, among which, as regards the agricultural sector, the most important is the introduction of new plant diseases in country where they were unknown before, so much so that experts even speak of "invasive alien species".

Global trade, movement of goods and products from one continent to another, tourism (the so called four T: trade, travel, transportation, tourism), allow as much as possible the spread of pests throughout the world. There have never been borders for plant diseases, after all (see box page 199). Yet, while in the past it used to take years for them to move from one country to another, now months, even weeks sometimes, are enough for a pathogen to easily cross borders and reach a new continent.

This acceleration is due to several factors. Goods travel with great speed. More and more farms producing plant material (seeds, bulbs, rhizomes, cuttings) have relocated their production to third countries, where climatic conditions are more favourable and labour costs are lower. Let's see some examples.

In the collective imagination, seeds production of vegetable and ornamental crops takes place in the Netherlands. However, since many years the main seed companies have moved seed production to tropical and subtropical areas: from there seeds are marketed simultaneously throughout the world. In addition to the possible spread of alien species, comes also the risk of phytosanitary problems due to the lack of specific analysis on the material that is exported and that might be a vehicle for pathogenic microorganisms and pests of different nature. To bear the brunt of this are importing countries, which have tried to protect themselves through the imposition of strict regulations aimed at preventing the introduction of the dreaded pests within areas where they are not yet present, or at least prevent their spread.

Several times, as part of my work, I observed the rapid and simultaneous spread of a pathogen in different geographical areas, due to the marketing of contaminated seeds. In fact, a very low percentage, even less than one out a thousand or even more, of contaminated seeds is enough to cause very serious yield losses in the field. It is something that has occurred many times and still occurs in the case of bulbs (tulips, daffodils, anemones, etc.) and leafy vegetables.

The European and Mediterranean Plant Protection Organization (EPPO) has been monitoring plant health for more than 50 years, and since the 1970s it has drawn up a list of quarantine pathogens which are not yet present in the Old continent and which therefore must be kept absolutely far from our European borders. A second list concerns organisms that are already present in some areas of our continents of which a further dissemination by means that vary from case to case has to be avoided. Lists are kept up to date and can be easily checked on the website at www.eppo.org; they inform about pests and pathogens which may, sooner or later, pose a serious risk to horticultural crops in many areas. Brother organizations, acting under the umbrella of the International Plant Protection Convention (IPPC), act in the different geographic areas (NAPPO in north America, APPPC in Asia and Pacific, NEPPO in the Near East... (see box page 200).

It is, moreover, pure utopia to think of being able to intercept a sick individual in large batches of seeds or of propagating material of the numerous species which are exchanged daily among different continents, despite all regulation in place (see box page 201). What is certainly important and indispensable is to further improve the rapid diagnosis of pathogenic organisms by developing an efficient network of laboratories. At the same time, it is essential to develop the means enabling pathogenic organisms to be rapidly detected and leading to a more rational and timely adoption of the available control measures, for promptness in the use of diseases management strategies is required at any time.

It is clear that globalization cannot be stopped and that concrete and practical measures must be adopted which are based on cooperation among different countries. This issue has become very crucial in the field of research and many international research groups are focusing on it in order to deepen all its aspects.

Going to depth

Transboundary plant pests

Plant pests have no borders and can easily spread everywhere. How do they travel? With goods (fruits, vegetables, seeds, cuttings etc.), by transport means, especially by plane (airplanes are excellent vectors of spores on the long distance) and with the passengers, who—more or less knowingly—may carry pests and pathogens from one country to another.

A real danger is represented by the many of us who, visiting distant countries, have the unfortunate habit of returning with seeds, flowers, seedlings etc.—material that can easily house dangerous pests typical of exotic countries and not yet present in our own countries. Some countries (such as Australia and New Zealand), which are highly geographically isolated, in order to prevent the arrival of new plant pests, are implementing very strict passenger control measures at any ports of entry.

Going to depth

Most feared pests

Among the most feared pests from overseas, there are *Phakospora pachyrhizi*, causal agent of soybean rust, able to cause yield losses up to 80%, and *Phytophthora ramorum*, a fungal pathogen affecting species of trees (especially oak), known for its devastating effects on the forests of western areas of United States and presently found in Europe only on ornamental species. Also other pathogens not reported in the list of EPPO (European Plant Protection Organization) or NAPPO or other organization might turn out highly harmful on species of remarkable economic importance for the different geographic areas. In the face of a clear regulatory framework on quarantine and its application measures, it should be noted that the strength of a country is represented by the adequate organization of its phytosanitary services, which are responsible for carrying out the controls—certainly not an easy task—and for taking the necessary measures to prevent introduction and spread of harmful organisms with a high phytosanitary risk.

Plant quarantine

Sometimes plant material (seeds, cuttings, plants) is quarantined, just as it happened many years ago to incoming ships, suspected of bringing people affected by contagious diseases, which were forced to avoid any contact on the land for 40 days to prevent the possible spread of illnesses. In plant pathology the term quarantine has a broader meaning, referring to the set of measures imposed by law and decrees, aimed at excluding from entire territories organisms or material that might carry plant pests still absent in those particular areas.

Quarantine may therefore consist in the total prohibition of a whole kind of products or plant material and in such a case it is justified especially in isolated countries or continents where pests coming from other territories would hardly have access, or banning or simply regulating access to particular products. The lack of timely and effective application of measures, even at a single entry point of goods from third countries can affect the neighbouring areas to the extent of spreading harmful organisms in the whole nation and into other neighbour countries, thus threatening the complex plant protection system.

Scary Spores

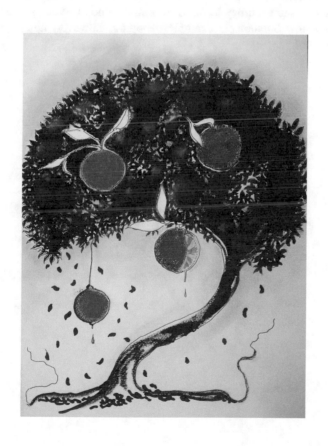

We are in the 2020s, the technology era. Yet so many plant pests and pathogens still scare us!

New plant diseases appear suddenly, devastating entire crops. They no longer cause famine and death, at least apparently. Thousands of growers, however, are forced to change their farming practices to deal with the damage caused by some plant diseases. Hundreds, thousands of hectares of crops are each time destroyed to eradicate a pathogen. The agricultural landscape can be affected and change forever. What happened in the Apulia region in Italy with the emergence of a new patovar of *Xylella fastidiosa* able to infect olive trees is just an example. And in the least industrialized countries, plant diseases continue to cause poverty and social hardship.

Moreover, there are pathogens used to destroy coca crops, and for agroterrorism. You can never lower your guard when it comes to pathogens!

Let us start a short journey to discover some of the most scary "spores" in the world—just a few but notable examples—hoping, however, not to frighten the reader too much!

Grapevine Flavescence Dorée

This grapevine disease is caused by a phytoplasma (a bacterium with no cell wall) transmitted by a leafhopper, *Scaphoideus titanus*, native to North America, introduced by accident to France after 1950 at a latitude around the 43rd parallel. Since then, flavescence dorée has spread to many other countries including Italy and attracted the attention of many researchers (see box page 206).

Its name comes from the pale green-golden yellow colour that the leaves of affected plants take on, resulting crumpled down and later reddish-coloured.

The twigs of affected plants fail to lignify, so they stay soft, hanging pendulously, and often die in winter as they do not resist the frost. The bunches of affected plants have few berries and do not ripen well. These symptoms can appear either quickly and lead to plant death, or at times, appear more slowly, and can even regress.

Chardonnay grape variety is among the most susceptible ones: flavescence has caused serious damage on it, worrying grapevine growers also for the lack of control measures against phytoplasmas. Entire vineyards can be wiped out by this disease, which has already destroyed many hectares of grapevine crops in north-western Italy, especially in Piedmont.

The use of insecticides against the vector insect proved to be effective in many situations, but it is not practicable in the case of organic vineyards. Another important aspect is related to the adoption of adequate plant health management at local level. A crucial role for the presence of *Scaphoideus titanus* is played by abandoned and uncultivated vineyards: it is here in fact that the insect can take refuge.

For some years there has been a decrease in flavescence dorée, with a partial remission of symptoms (see box page 207).

© The Author(s), under exclusive license to Springer Nature Switzerland AG 2021
M. L. Gullino, *Spores*,
https://doi.org/10.1007/978-3-030-69995-6_45

Double couple

Flavescence dorée only occurs if the pathogen-vector pair is present. Leading the fight against this strong couple is another winning team made of two researchers from Torino, Cristina Marzachì and Domenico Bosco, who are also a couple in life. Domenico, entomologist, studies the vector, while Cristina, a plant pathologist, studies the phyto-plasma. Together they have really studied all sort of things to eradicate this disease. Their commitment and passion in the management of this plant disease will certainly be rewarded.

Your symptoms will be "remitted"

In some plant diseases, sometimes a spontaneous remission of symptoms is observed, which in technical jargon is called "recovery". Almost like a remission of sins! It is right what occurred in Italy in the case of flavescence dorée: after an acute stage of infection, with severe symptoms, some plants started to produce symptom-free vegetation in the following season. Interesting phenomenon with considerable practical implications, recovery is a plant strategy that may be attributed, in addition to the mandatory control measures with insecticides, to the restoration of natural balances, according to which, after the outbreak of the epidemic, characterized by the sudden spread of a pathogen, spontaneously follows a phase of slowing down.

Sad Oranges

Citrus tristeza virus (CTV) is a very serious disease which affects several species of the plant genus *Citrus*, including sweet orange grafted onto sour orange rootstock, which was first infected in the 1930s, in Argentina and Brasil areas—a very serious epidemic that caused the death of almost a million citrus fruits a year. The name of this virus takes origin from the sad appearance and the general loss of vitality of infected trees. The major symptoms are seedling yellows, stem-pitting, reduced size of fruits and rapid death of plants.

The appearance of this disease is explained, in the one hand, with the spread of the pathogen, probably already present for a long time in Southeast Asia and the Malaysian archipelago, areas of origin of citrus fruits, and then exported to other countries, and on the other hand, with the massive use of sour orange as rootstock in order to avoid other pathogens' attacks. Citrus tristeza virus is now reported in all citrus fruit cultivation areas and, in Italy, where it has been present for a long time, threats southern citrus fruit production.

Because of its severity and the difficulty to manage it, this disease has deeply change citrus fruit industry worldwide. Farmers, in order to ensure an economically viable production, use certified virus-free plants and tolerant rootstocks instead of sour orange ones. But the replacement of sour orange, which is a universal rootstock for its ability of adapting to different pedoclimatic conditions, can penalize production both in quality and in quantity.

Closterovirus, the causal agent of the disease, spreads by aphid transmission but also by infected plant material, often imported without complying with quarantine standards (see page 201).

The legislation in many countries requires the use of plant material checked for CTV presence, but, unfortunately, in most citrus producing countries, non-certified nursery material is still produced and used. Trade globalization, together with the speed of transportation, also contributes to the spread of the virus. Monitoring the presence of the pathogen and its vector can be of great help, as confirmed by the experience gained by colleagues from Apulia (see box page 210).

© The Author(s), under exclusive license to Springer Nature Switzerland AG 2021   209
M. L. Gullino, *Spores*,
https://doi.org/10.1007/978-3-030-69995-6_46

Disease eradication has some effectiveness as a preventive measure but only in the short term. In Israel, for example, where this disease has caused very serious damages, plant material inspection, carried out on a very large scale using diagnostic techniques, has not achieved the desired results.

An interesting measure is the cross-protection technique, a kind of plant vaccination with attenuated virus strains.

A definitive solution for the management of citrus tristeza virus might come from the research and use of resistant rootstocks. Pending this, it is essential to implement disease surveillance programmes at territorial level, in order to reduce the risks of its spread, which might have serious consequences for the Mediterranean Basin, where citrus fruit production for fresh consumption is concentrated.

Going to depth

Virus and vector surveillance methods

Some Apulian colleagues, led by Donato Gallitelli, at the University of Bari, have gained interesting experiences in monitoring and prevention of Citrus tristeza virus, starting from the early detection of epidemic outbreaks to timely disease management in order to limit infection spread, and the eradication of affected plants and rootstocks. Eradication can be considered effective when the surveys carried out in the following three years give a negative result.

Fortunately, new monitoring technologies are now available, at territorial scale, such as GIS (Geographic Information System), advanced spatial analysis and remote sensing methods. Of course, monitoring must concern both the virus and its vector. Particular attention is paid in nurseries, because infected plant material is the main virus propagation channel in the free areas.

The Worries of Grasse Perfumers

It is said that it was Queen Victoria, in the 19th century, to launch the fashion of using dried lavender to perfume the linen. Certainly, this species is very popular and interesting because, thanks to its rusticity, can be cultivated in marginal yet sunny areas.

The world production of lavender amounts to 145 tons: 40% of it comes from France, the rest from Bulgaria, China and Moldova. Even in Italy lavender crops are spreading. Although it is certainly a "minor" crop, is still very important, interesting and peculiar of some territories.

As we know, Provence is characterized by the cultivation of lavender, whose pleasant and pervasive scent literally fills the air. Grasse is known for its lavender crops and finest essences (see box page 212).

Unfortunately, the lavender fields of this region, which have inspired generations of painters and photographers, have been recently endangered by a plant disease, whose causal agent, stolbur phytoplasma, is literally decimating these emblematic crops.

This disease is caused by a leafhopper-borne pathogen that completely invades the sap of the plant and within three years leads it to death. The phytoplasma attacks the roots of the plant, that wither and die; according to growers, between 2007 and 2010, *Lavandula x intermedia* crops—the most popular lavender species—used for soaps, perfumes and insect repellents, was halved by the new "plague". It was the dry summer and scarce rainfall to cause the insect vector proliferation. These leafhoppers lay their eggs several centimeters below the plants. The young insects limit to feed on the plants, while the adults also transmit the stolbur phytoplasma.

In France there are 1,700 companies working on about 16 thousand hectares of *Lavandula x intermedia*, along with other 4,000 hectares of lavender considered "pure"—an activity that employs 10 thousand people. The emergence of this phytoplasma and lavender crop decline have deeply worried both growers and perfumers.

Management of phytoplasma-infected plants is generally very difficult, especially in the case of a minor crop such as lavender. The use of insecticide to control

M. L. Gullino, *Spores*,
https://doi.org/10.1007/978-3-030-69995-6_47

the vector is not practicable due to the lack of chemicals registered on lavender. Therefore growers are focusing on alternative control means. Another possibility, though very expensive, is the use of greenhouse cultivation systems under protective plastic sheets that prevent the entry of vector insects.

And, above all, considering the economic value of this flowering crop, breeders are working hard in order to obtain resistant varieties. We all hope they will succeed, because lavender fields, in addition to making perfumers happy, are a feast for the eyes.

When Bruxelles sticks its nose into it

The perfume industry has been in a state of alarm and agitation for some years, due to the new European Regulation requiring further reduction in the use of natural ingredients, for their allergenic properties, in perfume production. Among the most controlled and regulated essences, there is also that of lavender, in addition to those of geranium and jasmine. This obviously scares high-end perfumers, because many of the most popular and appreciated perfumes contain these essences.

A Bacterium Destroys Kiwi Crops

When you say kiwi, you think of New Zealand. Instead, who would have thought that Italy has surpassed this country in the production of this fruit?

Kiwi (Actinidia) is native to Southeast Asia and in 1906 was introduced from China to New Zealand; from these islands, its cultivation has rapidly spread to the other countries. It was introduced in Italy at the end of the 1960s, interesting many fruit areas, from the north to the south, often previously devoted to apple and pear, converted to this species, as it is considered rather rustic, very productive and profitable. I remember very well when, at the beginning of the 1970s, many growers changed their cultivation to kiwifruit, while my father, as usually wise, chose to stay with peach and apple orchards, fearing that the new species would be too sensible to the low winter temperatures often reached at Saluzzo.

This fruit, which can be stored for several months, is very rich in vitamin C and has become very popular; for some years, close to the initially introduced species (*Actinidia deliciosa*), the golden kiwifruit cultivar (*Actinidia chinensis*) has started to be appreciated and marketed worldwide.

World production has exceeded 1.3 million tons; Italy has been the leading producer for some years. Recently, kiwifruit cultivation has become tricky because of the global spread of a bacterium, *Pseudomonas syringae* pv. *actinidiae* (PSA), which causes the typical bacterial canker lesions on canes and trunks, producing an exudate of varying colour from opaque white to dark red; the appearance of necrotic dark brown spots on leaves, often surrounded by a yellow halo; twig wilting and blossom necrosis with consequent premature fruit drop.

The bacterium was first observed in Japan in 1984 and then in South Korea (1992), in Italy (in Lazio area) in 1994 and in China (2004). Bacterial canker was the most serious kiwifruit disease, causing yield losses in many countries. In Italy, after the first appearance in 1994, the bacterium reappeared in 2007 in Lazio and has rapidly spread in all regions of kiwifruit cultivation, producing considerable damage and forcing the explant of several orchards.

Today all the kiwi-producing countries have been reached by this bacterial canker. The current population of the bacterium is able to multiply on both golden

M. L. Gullino, *Spores*,
https://doi.org/10.1007/978-3-030-69995-6_48

and green cultivars, while the strains of *Pseudomonas* responsible for previous epidemics were more adapted to green kiwifruit.

Since the causal agent of the disease is a bacterium that spread very easily, it is crucial to adopt preventive measures. In the case of orchards in areas not yet declared to be infected, parts of the affected plants should be carefully taken and sent to specialized diagnostic laboratories. Plants with conspicuous symptoms, severe decay and exudate, should be cut off and destroyed, at least as far as the aerial part is concerned.

In addition, when purchasing plants for new plantings, the production areas of the grafts and of the mother plants from which they originate should always be checked, to exclude the origin from infected plant nurseries.

Nursery production of actinidia has a great importance for protecting the industry. Moreover, in many cultivation areas it has been forbidden to replant kiwifruit in orchards previously explanted. As for all kind of pathogenic bacteria, also in the case of kiwifruit, disease control management is complicated by the scarcity of products with bactericidal activity and by the reduced availability of active ingredients registered for this crop (see box page 215). An interesting selection tolerant to bacterial canker has been found and selected (see box page 216).

The economic losses caused by bacterial canker of kiwifruit are very high and the development of the disease recalls that of previous severe epidemics (see box page 217).

Going to depth

Why antibiotics are banned in agriculture

Why in the 2020s is it so difficult, you may wonder, to counter a very small organism such as a bacterial pathogen affecting kiwifruit crops? Because, despite the many advances made by biology, today it is still very difficult to limit the development of bacteria attacking plants. You could actually do that by using antibiotics, just as you do for humans and animals.

However, in many countries, including Italy, the use of antibiotics in agriculture is forbidden. Why? For wise precautionary measures. As is well known, bacteria very easily select strains resistant to most commonly used antibiotics, making it difficult in some cases to protect animals and humans from certain diseases. Since the exchange of genetic information from one bacterium to another is quite easy and frequent, in many countries, the use of antibiotics in agriculture has been banned, for they would select very quickly resistant pathogen strains. With this precautionary measure, the passage of resistance characters from plant pathogenic bacteria to human and animal bacteria is avoided.

It is therefore a right and reasonable human health measure, although highly complicating plant protection from bacterial diseases.

Going to depth

Green Angel, a selection tolerant to bacterial canker

Genetic selection programmes have been carried out to search for tolerance and/or resistance to bacterial canker. A selection of *Actinidia chinensis* var *deliciosa,* "Green Angel®" is more tolerant to the disease than the varieties with green fruit used as comparison and when used as rootstock on red and yellow varieties, in comparison to "Hayward". "Green Angel®" induces a significant reduction in the percentage of leaves and leaf surface affected and the total absence of bacterial exudate on the trunk. The results obtained are interesting, because tolerant varieties can help to maintain the profitability of the crop by reducing the number of treatments. From the agro-pomological point of view, "Green Angel®" shows a reduced vegetative development, while the external characteristics of the fruit as well as the colour of the pulp does not show any differences compared to the reference cultivar Hayward.

Late blight of potato and bacterial canker of kiwifruit: is history repeating itself?

Can we see similarities between what happened in Ireland in the 1840s and what did happen a few years ago with bacterial canker of kiwifruit? Partly yes, partly no. Let's see why. In both cases, we are faced with a widely cultivated species and with the emergence of a new pathogen against which there are no fully effective control means. On the other hand, with kiwi, we are faced with an important yet not a staple crop. This means that the damage is certainly high from an economic point of view, but limited to the sector. In other words, hard times for growers and, to a lesser extent, for those who market kiwifruit, but no one will starve. Moreover, 170 years have not passed in vain and today, at least in the industrialized countries, the agricultural system is able to react faster to the arrival of a new adversity.

Xylella Destroys Olive Trees in Apulia

The news of the emergence in Italy of a new serious disease affecting olive trees in Apulia, leading them to rapid death, went around the world, along with the images of the area of Salento, literally devastated. The disease is caused by a bacterium, named *Xylella fastidiosa*, hitherto never reported in Italy, but sadly known from more than 150 years in America, where it causes "Pierce's disease" (by the name of the researcher who studied it), on grapevine and other species. This bacterium is particularly "demanding" from a nutritional point of view and therefore difficult to isolate with the techniques traditionally used in bacteriology. The attacks of this pathogen were known mainly in the United States of America, mostly in California and Florida, especially on grapevine, peach tree, plum, maple, oak, oleander and other arboreal species as well as on numerous herbaceous species; in South America (Brazil, Mexico, Venezuela and other countries) on citrus fruits, almond and other species. The introduction of *Xylella fastidiosa* in Europe occurred in an unpredictable way, and with a serious economic impact, as well as social, political and also environmental consequences. In some way the *Xylella* epidemic has similar aspects to the Covid-19 pandemic (see box page 220). Detected in 2013, the bacterium kept on spreading, from the initial outbreak, destroying millions of olive trees in Apulia. There are many evidences that this subspecies of *Xylella fastidiosa*, called *pauca*, has arrived in Italy with infected ornamental coffee plants from Costa Rica. Today, the spread of Apulian strain and/or further introductions of other strains could have a negative impact not only on the olive sector, but also for some key production systems of Italian and Mediterranean agriculture. Grapevine, citrus fruits, fruit trees, shrubs and ornamental and forest trees are at great risk as well as the entire nursery sector linked to them. This pathogen has been reported in other European countries, such as France and Spain, where it has been rapidly identified and confined. The bacterium, transmitted by an insect, a leafhopper, invades the woody vessels of the host plants blocking off their xylem, thus going to hinder the transport of water and causing foliar yellowing and wilting, with consequent rapid death of the plant (1–5 years after the attack). What are the main reasons for the rapid spread of the disease in Apulia? Certainly, on one hand, the ability of the bacterium to colonize many hosts (at least 150), typical of the

© The Author(s), under exclusive license to Springer Nature Switzerland AG 2021   219
M. L. Gullino, *Spores*,
https://doi.org/10.1007/978-3-030-69995-6_49

Mediterranean maquis (oleander, rosemary etc.) thus acquiring survival capacity, plays a role, and, on the other hand, the presence of a vector that allows the pathogen to move with a certain speed, behaving like a real hitchhiker.

The sudden appearance of this disease in Italy has encouraged intense research aimed at developing timely diagnostic methods. As with all bacterial diseases, management is really difficult, as antibiotics are banned in agriculture and also because reaching the pathogens within the xylem vessels is quite tricky. Suffice it to say that, despite the time that has passed since its appearance on grapevine in California, researcher did not manage to find effective means of control and eradication. In the meantime, they have started to test the presence of genetic resistance sources, in order to develop resistant cultivars. The first results obtained bode well for the future, but it is, of course, a very long job. Leccino cultivar, well known in Apulia, seems interesting for its tolerance to the bacterium.

For the time being, in order to limit the damage of *Xylella* on Italian olive tree crops, the attention has been focused on preventive and prophylactic measures, such as strict controls on propagation material (possibly infected scions) and infected plants, with consequent elimination and burning of plants or parts of them, as well as on the maintenance of good nutritional conditions and a healthy vegetative balance, factors that may hinder the infection.

Once again, 180 years after the Irish famine engendered by late blight of potato, despite the progress made in the meantime, a new pathogen shows us the vulnerability of the agricultural system, associated to deaths and famines in the past, to huge economic and environmental damage today, particularly to the olive oil industry (see box page 221).

Covid 19 and *Xylella*: so many similarities!

Epidemics, whether concerning humans, animal or plants, all resemble each other. In fact, between the epidemic of Covid-19 spread worldwide, of which we have been suffering the consequences for a long time, and that of *Xylella* on Apulian olive trees, there are many similarities, ranging from the significant delay in pathogen detection, to subsequent difficulty to stop the outbreaks once they started to be too many at the same time.

Extra virgin… almost pregnant

Long before it came fashionable to fight food counter-feiting in the early 1980s, living in the United States and going to the American malls, I realized how many manufacturers tried to cheat customers playing with assonances and making them believe they were buying real Italian products. I was very impressed by a very low quality Spanish oil called Italian: many Americans bought it convinced to consume an Italian product.

Moreover, the average American was not yet pre-pared to understand the delicacy of true extra-virgin olive oil. I still remember the popular joke of that time in USA: "Extra virgin? Almost pregnant!".

When the Green Gold of Liguria Gets Sick

The fragrant basil (*Ocimum basilicum*) is a real ambassador of Liguria in the world. Liguria region is in fact famous all over the world not only for its blue sea but also for this culinary herb, with a delicate scent, which is either consumed fresh cut or transformed into the famous pesto, and which grows well at sea level. And if the world is full of other "basilica"—whose leaves can be red, larger and more or less aromatic, often inedible, and even grown at remarkable altitudes—the one everybody refer to as basil par excellence or, better, "His majesty the basil", is the sweet basil of Liguria. Thanks to the selection of varieties characterized by an incomparable aroma, basil is for Liguria more precious than gold, so much to be called "green gold". The name of this highly aromatic plant comes from the ancient Greek term *basilikon*, meaning "royal", perhaps because it was used in the embalming processes as a good omen for pharaohs' afterlife in the ancient Egypt or for royal ointments in Persia. Referred to as *basilicus*, in ancient times it was considered useful to fight dragons or, more simply, a good remedy for reptile bites. Moreover, still today in India, its antiseptic properties are exploited against insect bites. And if the place of origin is unknown and probably not unique (tropical Asia and India in particular, but also tropical Africa), certainly Liguria has become the preferred land for the cultivation of this species. Traces of its discovery in Provence and Liguria were found in the diaries of the Romanian and Jacopei pilgrims in the mid-fourteenth century. For Liguria, basil represents a high value crop. Grown in open field for industrial production, and in protected cultivation systems—where it has smaller development, a thinner stem, a lighter colour and a more delicate aroma —for fresh consumption, basil is an interesting source of income for farmers.

No wonder, the care that is given to its cultivation. The fragility of this crop, the frequent harvesting, the presence of several pests, especially fungi, capable of causing serious losses, require in fact the adoption of sustainable management strategies.

Only a few years ago, a dangerous disease, downy mildew, threatened Ligurian basil crops, so much to fear for their survival. For a few months, at the beginning of the years 2000s, it was impossible to eat pasta with pesto in Ligurian restaurants.

When the customers ordered this dish, a restaurateur from Sanremo answered in despair: "Gentlemen, we will never be able to make pesto again (see box page 225). A serious plant disease has attacked basil-growing areas—a kind of SIDA. All the greenhouses crops have been affected!"

What ever happened? In the fall of 2003 a new basil disease suddenly appeared in Ligurian greenhouses. A mould covered the leaves that, once reached the consumer, rotted within 1-2 days.

In a region where 100 hectares are devoted to basil crop, the emergence of this new disease led to huge yield losses and redundancies of many people.

Within a few weeks, the mysterious disease was identified as basil downy mildew, a highly contagious and devastating disease caused by a fungus-like organism with a very complicated name, *Peronospora belbahrii*.

Indeed, basil downy mildew attacks had already been observed in the 1930s in Uganda, but the distant report of it remained for years at the level of curiosity, until 2002, when the disease reappeared in a family garden in Switzerland. Much more serious, however, from the economic point of view, were the attacks observed immediately afterwards in Italy, in Liguria and Piedmont, since the autumn of 2003.

The disease then spread very rapidly in all basil-growing areas, starting from the nearby Côte d'Azur, in France, to reach California. Why so quickly? Because pathogen transmission, as proved by the researchers of Agroinnova, occurs via contaminated seeds. In fact, a very low percentage of seeds, of the order of 0.02%, is sufficient to ensure a very rapid disease spread (see box page 226).

In 2014 basil downy mildew made the news again in Italy, so much the newscasts opened with updates on the basil "plague".

But let's stay in Ligurian basil crop areas, such as Genova, Savona and Imperia. From Genova to Diano Marina, the 80% of farms have been affected by this disease, with huge damage for the local economy. The restaurateur from Sanremo, although in a colourful and perhaps exaggerating manner, had certainly centered the problem: basil growers immediately went into crisis, there were layoffs and it took a few years to at least partially repair the damage caused by basil downy mildew.

Basil is, in fact, a so-called "minor" crop, because, although very important in Liguria, is present worldwide just on limited areas, therefore at the time of basil downy mildew epidemics, growers could not count on the presence of registered fungicides for its management. Today basil mildew is still a serious and dangerous disease but, thanks to the efforts of researchers, farmers can count on a variety of control measures (chemical, agronomic...) of a certain effectiveness (see box page 226).

One hundred and sixty years after potato late blight, another mildew has come to afflict a crop, basil this time (see box page 227).

Pesto for pasta

Curiosity

The American entrepreneurial spirit took me so much that at a certain point of my stay at Cornell University, in 1989, I started with Judy Hall (called Judy Mall because of her shopping-addiction), a fellow PhD student at the Department where I worked, a semi-industrial production of pesto. "Pesto for pasta" was our brand. We grew "Genovese gigante" basil with the seeds I brought from Italy and we used Parmesan cheese with which I packed my suitcase every trip back from Italy. We quickly filled a freezer with pesto. Terry Acree, my landlord and famous professor of food technology at Cornell University, did all the necessary tests to allow us bring our pesto to supermarkets. On Sundays, Judy and I went to Rochester's Safeway (the most exclusive supermarket chain in the United States) and set up a tasting-table of our specialty. I admit that at some point I seriously considered the idea to start an entrepreneurial business. I already saw myself, as Diane Keaton in *Baby Boom*, producing thousands of pesto pots! But, of course, my passion for research was too strong and I wouldn't change direction. I can't imagine what my Torino professors would have said if I had stayed in the USA to produce pesto. Come to think of it, this was my second spin-off. The first one being the one started with rotten apples in my teens (see page 127).

Curiosity

Beware of seeds

The sudden arrival of basil downy mildew and its rapid
spread in several production areas was explained by seed
transmission of *Peronospera belbarii*. In this way, the
seeds, produced in a few plants in tropical areas and
marketed all over the world, represent formidable
vectors. The question is whether the seed industries
are aware of the danger arising from seed production in
geographical areas where the disease is endemic and
there is the enormous risk of spreading it with seeds.

A chief doctor at the basil bedside

It took a few months for researchers and local
authorities to organise themselves in such a way as
to enable farmers to have adequate means of control.
At the same time, rapid diagnostic methods and seed
treatments were developed to verify seeds' health
and decontaminate them if needed. Hence the
presence of real "doctors of basil" able to attentively
take care of "His majesty's crops"! In the case of
basil downy mildew, it was a real "chief doctor", the
Ligurian Angelo Garibaldi, professor emeritus at the
University of Torino, now also known as
basil's "savior".

Potato late blight and basil downy mildew: a comparison

Basil downy mildew appeared as quickly as did potato late blight, but in a very different context. First of all, although of great importance in some geographical areas, basil is not a staple crop as was potato in Ireland at the time of potato blight epidemics. The damage, therefore, has been purely economic for agricultural companies. Moreover, today we are able to counteract much more quickly to the emergence of a new plant disease. In other words, the damage was high, but limited to an economic sector. The epidemiologic principles, though, were and are the same.

The "Silly" Plant that Troubles Rice Growers

A fungus with a rather difficult name (*Fusarium fujikuroi*) has been attacking rice seedlings for years. Found for the first time in Japan in the far 1828, it is now widespread in all rice-growing areas, including Italy.

From seeds infected by this fungal pathogen originate rice plants that show an anomalous development, lengthening much, blooming precociously and producing sterile panicles. These plants, higher than the healthy ones, stand out in rice crops. This anomalous and exagerated development has been explained, since long time, with the production by the pathogen of gibberellins, which are nothing less than plant growth hormones.

The disease caused by this fungus is called "bakanae" which in Japanese means "silly, ridiculous plant", precisely because of the weird aspect assumed by affected plants.

Known since 1938, bakanae had never been a serious problem in Italy until the beginning of the 1990s, when it showed an increase of severity in rice-growing areas.

Why? On the one hand, because all varieties of Italian rice are susceptible to this pathogen, and on the other hand, because the fungicides used as control means for other diseases are not very effective on it. Besides, perhaps bakanae has been a little underestimated. These factors, together, have come to worry Italian rice growers, and not only them.

Fortunately, this pathogen, like all those seed-transmitted, can be managed by treating seeds either with chemical seed-dressing (carried out with fungicides) or with thermotherapy, consisting in water immersion at 60 °C for 15 min or aerated-steam disinfection at 74 °C for 2 min. This latter treatment technique is very effective, leaves no residues and has a very low environmental impact, apart from energy consumption.

Hot water seed treatment for fungal diseases is already used by some seed companies with excellent results. This kind of treatment, however, has the disadvantage of being able to be applied on small quantities of seeds, while the aerated steam therapy, developed within European projects (see box page 230) is not yet applicable in a large scale as the machinery used for industrial scale treatment is very expensive and not for

© The Author(s), under exclusive license to Springer Nature Switzerland AG 2021
M. L. Gullino, *Spores*,
https://doi.org/10.1007/978-3-030-69995-6_51

sale, at least in Italy. A real pity, since Italian rice growers, who are producing the very precious "Carnaroli" rice indeed would need it! (see box page 231).

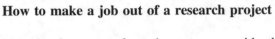

How to make a job out of a research project

The development of seed treatments with thermotherapy (on cereal and vegetable crops) has been the subject of two major project funded by the European Commission under the Fifth and Sixth Framework Programmes. A group of Swedish researchers, as part of these projects, has developed a machine for seed treatment with aerated steam which allows to obtain excellent results in terms of effectiveness against several important pests, without damaging seed vitality. The Swedish research team included a brilliant student who, after completing his Ph.D. thesis on this subject, made of seed treatment with this machine his job.

Carnaroli: the best risotto rice is produced in Italy

Carnaroli is an Italian medium-grained rice grown in the provinces of Pavia, Novara and Vercelli in northern Italy. Carnaroli is used for making risotto, differing from the more common arborio rice due to its higher starch content and firmer texture, as well as having a longer grain. Carnaroli rice keeps its shape better than other forms of rice during the slow cooking required for making risotto due to its high amylose content. It is the most widely used rice in Italian cuisine, and is highly prized.

Scary Threats Causing Bellyache

Fresh fruits and vegetables may be contaminated by microorganisms potentially pathogenic to humans, such as, to name but a few, *Salmonella* spp., *Escherichia coli*, *Staphylococcus aureus*, *Clostridium* spp. and *Listeria monocytogenes*. It is mainly *Salmonella enterica* and *Escherichia coli*, producer of Shiga-toxin (STEC), that are responsible for recent outbreaks associated with fresh-cut produce consumption.

In Europe, general public attention towards this phenomenon, also due to the high number of deaths, was prompted by the episode of the so-called "killer bacterium", which occurred in Germany in the spring–summer 2011 (see box page 235). In fact, the German case followed outbreaks of *E. coli* or salmonella associated with sprout consumption, as also occurred in the United States, Canada, Japan and other European countries (see box page 236).

The presence of *E. coli* or *Salmonella enterica* on fruit and vegetables is therefore an important food safety problem, also because of the excellent survival capacities demonstrated by these microorganisms. But why do plant pathologists care about this phenomenon that seems to be of exclusively microbiological nature? Because some strains of human pathogens have proved to be able to colonize plants with mechanisms sometimes similar to those used by plant pathogens themselves.

Soil fertilisation and irrigation techniques may favour plant contamination. Since enteric bacterial flora of farm animals may be colonized by human pathogens in absence of symptoms, cattle faeces are the main source of these pathogenic organisms. Irrigation water taken from areas close to farms or pastures may be contaminated by animal waste, sewage and livestock waste.

The application of manure and irrigation water contaminated by *E. coli* 0157:H7, a strain sadly known for problems caused in different geographical areas, may allow the bacterium to reach the tissues of lettuce or other fruit and vegetable products through the root system. The irrigation technique used may affect plant contamination: surface irrigation favours bacterium spread, compared to drip irrigation.

Contamination in the field can occur directly on the edible parts or on other parts of the plant, from which bacteria then reach the tissues used for feeding. Using plant

© The Author(s), under exclusive license to Springer Nature Switzerland AG 2021   233
M. L. Gullino, *Spores*,
https://doi.org/10.1007/978-3-030-69995-6_52

as a host, the bacterial species survives in the hostile environment adapting to new physico-chemical conditions; the colonization of plant tissues allows the bacterium to enter the food chain and reach the intestinal microflora of the host (animal or man) through ingestion of vegetables. Human pathogens may be present on the host superficially, or, more subtly, in the internal plant tissues, affecting leaves, roots, flower and even seed. The presence on the seed allows pathogens to be transmitted from one plant generation to another, ensuring the species survival and therefore increasing their dangerousness. The seeds of some vegetables, as already seen, are used for the production of buds; if they are contaminated by pathogenic bacteria, even the buds that are obtained will be contaminated.

The ability shown by some bacteria not only to survive, but also to colonize fresh vegetable crops makes it necessary to develop control strategies complementary to those used by the agri-food industry during produce processing and storage stages.

The use of agronomic techniques, in particular fertilisation and irrigation, which limit the spread of bacteria on fruit and vegetables is of primary importance for farms supplying the ready-to-eat sector. If contamination of agricultural products is low, treatments carried out in processing and packaging facilities will be less extensive and more effective.

Precisely in view of the ability of certain bacteria pathogenic to humans and animals to colonize plants, it is interesting to evaluate the effectiveness of chemical, physical and/or biological control means currently used for fungal and bacterial disease management, always in accordance with the current legislation.

Research is also moving towards the selection of microorganisms able to counter the development of human pathogens (see box page 237). Also in this area, it is essential to rely on effective detection methods and rapid diagnostic tools.

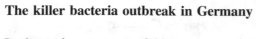

Going to depth

The killer bacteria outbreak in Germany

In the spring–summer of 2011 an extensive epidemic occurred in Germany caused by a new strain of *Escherichia coli* (O104:H4,) capable of producing Shiga toxin. The number of cases was impressive: over 4,000 cases of hemorragic diarrhea, of which more than 900 with Hemolytic Uremic Syndrome (HUS), a condition that clogs the filtering system in the kidneys leading to renal failure, necessitating dialysis treatment, and over 50 deaths, especially in Lower Saxony. Also people of other countries, having travelled to Germany during the epidemic period and consumed contaminated food, were affected.

The first microbiological investigations erroneously indicated the possible source of the epidemic in the consumption of cucumbers imported from Spain. More in-depth epidemiological searches led to the identification of the outbreak source in the consumption of raw vegetable sprouts. All products involved in the outbreak came from a company in Lower Saxony producing sprouts from various seeds: fenugreek seeds, purchased in Egypt, were responsible for the infection.

All in all, a bad, poorly managed story that caused panic among consumers and greatly damaged Spanish agriculture in particular and European agriculture in general!

Troubles not only in Germany

German case followed sprout-related disease *Escherichia coli* or salmonella outbreaks occurred in the United States, Canada, Japan and other European countries. Well known for a long time are the many cases of fresh vegetables contamination by *E. coli* O157:H7 occurred in the United States of America.

Before the killer bacteria outbreak in Germany, several cases of vegetable contamination by *E. coli* O157:H7 and subsequent epidemics were reported in Europe, for example in Spain, Ireland and Sweden. Often a contaminated product, if exported to more than one country, may determine in the areas of destination, diseases caused by the presence of microorganisms in the intestinal mucosa due to ingestion of contaminated foodstuffs and to the presence of toxins produced by the microorganisms themselves.

Going to depth

Trends in the search of alternative disease control measures

Since, as already seen in the case of bacterial canker of kiwifruit, apart from copper products, there are no pesticides with bacterial action registered for agricultural use, it may be interesting to evaluate the possibility of exploiting the antagonism by certain microorganisms, such as bacteria and yeasts, against human pathogens, especially during pre and post-harvest stages.

Some bacteria, isolated from alfalfa sprouts, are able to counteract contaminations by *Salmonella* spp. Even some lactic bacteria and yeasts, under experimental conditions, show an interesting activity against human pathogens. Another group of antagonistic microorganisms tested against human pathogens is that of yeasts. Therefore, the application of antagonistic microorganisms might assure the safety of vegetable produce in the future, in absence of other control methods.

Also the application of essential oils (for example oregano or basil oil) or compounds extracted from organic matrices, or microorganisms, in association with other techniques to be used during processing, might be a valuable tool for improving horticultural products' safety.

Fungal Troops Against Narcotic Plants

Fungi can be used as a means to damage drug plantations: selected pathogenic species can be exploited to target illicit crops. This is the one of the last frontiers in the fight against drug trafficking.

The idea of using fungi as herbicides against marijuana, coca and opium poppy crops comes from far away: it is born in the United States.

Several studies, some of which carried out by colleagues of the National Research Council of Bari, have proved the potential effectiveness of selected fungi in killing narcotic plants: *Brachycludium papaveris* and *Crivellia papaveracea*, for example, are effective against opium poppy. *Fusarium oxysporum* f. sp. *cannabis* affects cannabis, *F. oxysporum* f. sp. *ferythroxyli* is pathogenic on coca. Most of these fungi produce leaf necrosis or attack host vascular tissues causing wilting, loss of turgor and finally death of the plant.

They are fighters employed to attack coca crops and combat the international drug trafficking.

But is it really a viable undertaking? To answer this question, a study project has been carried out in which all the present skills and the data on effectiveness, safety, feasibility and large scale application of pathogenic fungi collected so far have been lined up together with the subsequent steps to put in practice to achieve this ambitious goal.

From a scientific point of view, the use of fungi against narcotic plants has a high potential and is a valid drug-fighting means, but the research carried out so far is not sufficient for its practical application.

Fungal pathogens of drug crops are not yet used to fight the 600,000 hectares of cannabis grown mainly in Afghanistan, the 200,000 hectares of coca crops concentrated in the Andean areas of Peru, Colombia and Bolivia, and the 130,000 hectares of opium poppy production in Afghanistan, Mexico and Myanmar (see box page 240).

The reason lies in the difficulties encountered in their practical application. Although these pathogenic fungi are well manageable in laboratory, stable over time, addressable to precise targets and adaptable for industrial production, to date their environmental impact has not been fully evaluated.

© The Author(s), under exclusive license to Springer Nature Switzerland AG 2021   239
M. L. Gullino, *Spores*,
https://doi.org/10.1007/978-3-030-69995-6_53

Many of them destroy the drug crops with toxic substances: *Fusarium* species, for example, produce a wide range of toxins, which can invade tissues of target plants and, after killing them, have a dangerous impact on other organisms that share the same environment, such as other plants or animals, including human beings. Therefore, further research is needed to understand whether fungi used against drug crops might somehow endanger the environment. Surely there is a discrepancy between the potential of these new weapons against drug trafficking and the absolute not easy context in which they would be used.

Difficult countries

Another aspect to be taken into account is, moreover, the complexity of drug cultivation countries. Take for example Colombia, a country where, by means of international agreements, US military interventions and aerial raids with herbicides are tolerated in order to limit drug trafficking, but no authorization is ever given to carry out tests for the use of pathogenic fungi against drug crops.

Moreover, after visiting Colombia and its coca crops in 1997, I realized how difficult it would be in certain areas to grow other crops, in the absence of infrastructure. Absolutely normal families, living in beautiful landscape, but completely isolated, make a living by growing coca so to be able to buy shoes for their children and send them to school. If they ever grew apples instead of coca, who would they sell them to?

Other nations are no less troubled. Many of them are at war and therefore hardly reachable. Those that are not, on the other hand, have to contend with farmers, who do not have alternative income than that coming from drug cultivation: any other legal crops would not have buyers. Caught in the grip of an illicit trade, they have to sell their produce for a few dollars to organizations that will earn a thousand times as much.

Agroterrorism

Bioterrorism is a very important issue for the international community. Traditionally, the term has been associated with risks arising from infectious diseases, genetically modified organisms and biological weapons. In a broader sense, however, it is used to indicate prevention from biological risks arising from intentional and unintentional events, not least the deliberate introduction (as well as the threat of use) of plant pests able of endangering a country's agricultural economy, production quality and safety, consumer confidence and, more generally, the national welfare.

In a general situation where more and more attention is paid to prevention of terrorist attacks towards high value targets, also agriculture—together with the agri-food system —is now endangered by agroterrorism, enhanced by individuals or groups, with the aim either of causing direct damage to crops and forests, or indirectly influencing the agri-food sector which, with related activities, is crucial to social, economic and political stability of any country.

For some years now, a growing number of researchers have focused their attention on the issue of agroterrorism. Why?

The many terrorist attacks, starting from the one in New York City in 2001, those of 2005 in Spain and Great Britain, and 2015 and 2016 in France have undoubtedly left an indelible mark on our society, bringing back old fears and making us feel more fragile and exposed to violent and destabilizing actions, once considered impossible or, at least, highly unlikely.

In Europe, even if only 5% of its population work in the agricultural sector, any terrorist actions would lead to dramatic consequences not only in the sectors of productions, processing and trade, but also in the related catering industry and tourism. Not to mention what an agroterrorist attack would mean in a developing country where, in the lack of means and structures capable of identifying in a short time any agents, deliberately introduced, responsible for serious phytopathological issues, the situation would add to an already critical technical and cultural background.

M. L. Gullino, *Spores*,
https://doi.org/10.1007/978-3-030-69995-6_54

Yet, might plant pests be used as biological weapons to destroy entire crops or forests? Or contaminate agricultural commodities? The idea is unfortunately neither new nor absurd and it suffice to look back into history to realize that the hypothesis of deliberate introduction of plant disease agents is not an absolute novelty (see page 65).

Have biological weapons programs really been abandoned nowadays? Probably not. In 1995, for example, the UN Special Commission on Iraq highlighted an activity in that country to produce biological weapons to be used against neighbouring states. Iraq began pursuing offensive biological warfare in the 1970s, the more intensively in the years 1985–1991, carrying out research on the production of pathogenic fungi of cereals, especially of *Tilletia caries*, causal agent of common bunt of wheat. Iran's wheat crops were the main target of Iraq's research in the 1980s–1990s. In this case, the aim was spreading the agent of what was an insignificant disease in the industrialized world, but still alarming in the developing countries.

Another potential risk, much more perceived by consumers, is the possibility of food being contaminated, for example with pesticides or, more subtly, with toxin-producing agents. And in fact we can find many examples of this kind! (See box page 244).

Agricultural crops, forests and agri-food products of the industrialized countries are therefore a potential target for terrorism. Why? On the one hand, they neither are or can be protected in any way and, on the other hand, the use of pathogens against them requires relatively simple, inexpensive and easily accessible technologies. The fact that plant material is increasingly being produced in very few plants and distributed around the world encourages, moreover, as we have already said, the rapid spread of plant pathogens. It is easy to figure out the possible consequences of distributing batches of contaminated seeds in different geographical areas, at levels that cannot be easily detected by plant protection services, but sufficient to spread epidemics in different geographical areas. As we have already said, even one out of 10,000 infected seeds can seriously compromise a harvest.

Another risk is the possible negative effect on the export of certain products, caused by their possible exposure to pathogens of dreaded introduction: the so-called quarantine pathogens.

The economic importance of some crops worldwide could, therefore, justify any agroterrorist attacks, which has become an interesting topic of research (see box page 245).

At the same time, recent episodes (just think of the energy blackouts occurred in the United States in August 2013 and in Italy in September of the same year), clearly give an idea of the similar but much higher and faster impact that further actions might have on sensitive targets—such as electricity supply service, drinking water supply etc.—if carried out for terrorist purposes.

The real chance of successfully using plant pathogens and other pests as biological weapons arouses a number of technical and ethical concerns. Skeptic are those, among plant pathologists, who have more experience and familiarity with field trials. We all know, in fact, how difficult and complex can be carrying out artificial inoculations in open field. Fungal and bacterial pathogens are generally strongly influenced by environmental factors, such as temperature, relative humidity and light. Fungal spores, moreover, only in some cases have the ability to spread over great distances.

Even under the most sophisticated experimental conditions, not all artificial inoculations result into infection. Even more unlikely is to be sure that a pathogen, once introduced in large quantities into the environment as a biological weapon, can really cause the appearance of the disease with the desired severity.

The political context arisen in the years after 9/11, 2001 obviously justifies the fact that a previously far-fetched occurrence such as biological warfare might become reality.

New plant pests are continuously introduced into our countries and reach our crops, often through the commercialization of propagation material. Prevention measures and strategies against deliberate pest attacks are also useful to counteract the "natural" and very frequent pest introduction.

Moreover, the current situation of insecurity means that the deliberate use of dangerous plant disease agents on high-income crops cannot be ruled out. Today the risk of agroterrorist attacks cannot be excluded and, at least in certain countries, it is carefully taken into account. In fact, it has already influenced the relations that some of us have with foreign colleagues, especially Americans, making it more complicated to exchange plant material and fungal cultures in particular.

Certainly, in the case of an agroterrorist attack, industrialized countries have structures and means capable of timely identifying any deliberately introduced agents responsible for phytopathological issues, therefore implementing the necessary control measures. Developing countries are more vulnerable, as any biological warfare against them would exacerbate already critical technical and cultural situations.

Stories of bellyache

In the 1970s–1980s several terrorist groups repeatedly threatened to contaminate food batches aimed to European or North American market. The harassment of Tamil Tigers (more or less clearly supported by the Indian government) is perhaps the most documented case of the 21th century. In 1952, however, in Kenya an organization of Mau Mau rebels killed 34 cattle for fattening using a toxin of plant origin. In 1984 a bioterroristic attack carried in Oregon by a religious organization caused the food poisoning of 751 individuals through the deliberate contamination of salad bars with *Salmonella typhimurium* strains at ten local restaurants. In 1996, again in the United States, a fired employee caused the intoxication of 12 of his former co-workers through intentional food contamination in the corporate office's cafeteria with *Shighella dysenteria* from a laboratory's stock strain. In these last two cases, the blameless victims got away with a good bellyache.

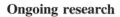Going to depth

Ongoing research

In some countries, such as the United States, the subject of agroterrorism has been given the utmost consideration in the last 25 years and, with the typical pragmatism, Americans suggest a practical approach aimed at balancing two strategies: prevention and preparedness. While the US Congress has focused more on a preventive approach, issuing a series of new laws aimed at increasing national security against the risk of offensive biological attacks, community attention has been more focused on the importance of promptness, in the form of development and availability of techniques for the rapid recognition, diagnosis and control of any plant diseases introduced as biological weapons. To implement this, the United States Department of Agriculture (USDA), has strongly reinforced the diagnostic services, strengthening existing laboratories, opening new ones and networking them. A planetary diagnostic system, capable of predicting the onset of epidemics (like a weather forecast) seems to be the next challenge for global biosecurity. American colleagues, with their concreteness, have established priorities for many research activities. A strong boost has also been given to research projects concerning the epidemiology of pathogens considered at risk of introduction as terrorist weapons. Research topics that had not been investigated before, such as microbial forensics, are becoming increasingly important, even in relation to biosecurity. Great importance is also attached to the evaluation of the ethical aspects involved in this particular field.

In Europe, the subject of agroterrorism, initially considered remote, has begun to gain attention since the 2000s. The Community shows great determination in networking in order to control the spread of certain pathogens. The American commitment has certainly motivated my research on this topic, which I find fascinating. A group of European researchers with experience in different branches of plant pathology (from diagnostics to epidemiology), recognising the potential risks posed to European agriculture by the

Going to depth

deliberate introduction of plant diseases, established, with the coordination of Agroinnova throughout various European projects, a network to validate the technical performance of diagnostics in the phytosanitary field and jointly plan effective strategies for monitoring and preventing pest introduction and spread.

Spores of the Future

I don't presume to read the future, but in this part of the book I want to look ahead, trying to imagine future scenarios, with the awareness that today changes are so rapid that sometimes we find it difficult not only to adapt to them but also to imagine them. On the one hand, I have addressed some of the issues that are likely to affect plants and their diseases in the coming years; on the other hand, I have tried to imagine how the plants will be treated in the future. SPA for stressed out plants? Hypnosis to heal plants? How will tomorrow's plant pathologists be? The recent Covid-19 pandemic indeed disrupted everything, changing the game, not only in human health. Also our discipline, its role, its importance and social impact will change. Let's start this last part of our trip!

Climate Change and Future Scenarios

Almost everyone now agrees that the climate is changing and that is not so easy to influence this process. Industrialized countries, in fact, continue to consume, while the emerging and developing countries, whether we like it or not, are eager to conquer all the well-being (or presumed so) we have been enjoying and exploiting for years.

As already seen (see page 59), climate change affects plants and their diseases in a manner bound to worsen, at least in the next ten-twenty years.

The possible consequences, consisting in yield losses due to plant diseases, effectiveness impairment of control strategies and variations in the geographical distribution of pathogens are now yet completely known, despite much research is currently addressing this issue.

Global warming is causing a general shift of the agroclimatic zones and associated crops towards the poles. At the same time, pathogens will follow their hosts in this migration, spreading and adapting themselves to the physiological and ecological adaptation of the plants to the new environment.

Changes in the distribution and incidence of plant diseases might threaten the economic competitiveness of the countries affected by these phenomena in two ways. First, diseases have a significant impact on production costs; second, the expansion of pathogen diffusion ranges and the possible increase of mycotoxin production (induced by more frequent and abundant precipitation) might lead to restrictions on the export of plant products. The possibility of introducing new pathogens and the presence of mycotoxins in food represent a threat and a risk to human and animal health, which is so serious that it justifies an intensification of barriers to international trade, implemented through the revision of phytosanitary regulations.

In addition, the emergence of new or more serious phytosanitary issues might lead to change or abandon of seed production in areas previously relatively plant disease-free.

The literature available on plant pathology and climate change deals with experimental research carried out with the aim of investigating the effect of any

© The Author(s), under exclusive license to Springer Nature Switzerland AG 2021 249
M. L. Gullino, *Spores*,
https://doi.org/10.1007/978-3-030-69995-6_55

variation in the composition of atmosphere on the combination of several hosts and pathogens and assessing future scenarios developed with climate simulation models (see box page 251).

The expected temperature increase seems likely to cause an expansion of the geographical distribution of pathogens towards the poles and an increase in the number of generation per year. It is also assumed that rising temperatures will increase survival and hence the initial amount of inoculum, while the increase of drought in continental areas during summer might reduce the risk of infection by those pathogens that need foliar moisture or soil saturation as environmental conditions for infecting plants.

Whether such assessments can reflect the future is difficult to predict, as the phenomena considered arise from the complex interaction of atmospheric, climatic and biological factors with technological and socio-economic changes, in turn extremely critical to predict. However, the use of simulation models is, to date, the most suitable and economical tool to understand how the different combinations plant-pathogen (defined pathosystems) can react to climate change. The estimates obtained should be assessed by taking into account the margin of error inherent in the modelling of highly variable phenomena such as biological factors and the amplification that this margin may suffer due to the integrated use of different models.

Since it is not easy to reverse the trend, mitigation interventions are very important, aimed at reducing the negative impact of climate change. In agriculture, there will be strong changes in cultivation patterns and new varieties will be selected that are more suitable for tolerating higher temperatures.

Some countries, especially in Africa, will be characterized by migrations of entire populations, who will move in search of water and lands to cultivate. Climate migration is undoubtedly among the growing phenomena of the 21st century.

Going to depth

Future scenarios

The effects of climate change on plant pathogens taken as a whole are evaluated through future scenarios resulting from the coordinated use of climate simulation, cultivation and epidemic models, in order to predict the potential trend of infections and the consequent yield losses. Climatic mapping programmes, in the meantime, provide a visual representation of spread risk, allowing the identification of geographical areas potentially susceptible of infection —from a climate point of view.

The combination of such programmes with other climate simulations would also allow to predict pathogens' invasion and so to assess the vulnerability of an ecosystem or region not only on the base of a climatic projection but also on the characteristics of pathogens, host plants and environment. The outputs obtained from these instruments, therefore, might be used in order to develop adequate quarantine measures to limit the risk of import of plant material from infected areas.

In addition, the likely absence of direct effects of increased atmospheric carbon concentration on fungal pathogens makes climate mapping programmes particularly suitable for predicting the effects of climate on pathogen distribution and disease severity. In fact, despite the indirect influence of CO_2 enrichment through changes in the host physiology and morphology, the variations in thermo-pluviometric regimes—main inputs of epidemic prediction models—have the greatest influence on pathogen epidemiology.

Green Biotechnologies

The use of biotechnology is well accepted in the medical field (the so called red biotechnology) as considered useful to our health, and tolerated, perhaps because not largely known, in the industrial field (white biotechnology), while it raises great concerns in the case of plants (green biotechnology).

The development and use of genetically modified organisms (GMOs) and, in particular, genetically modified plants (GMPs) is a subject of intense debate in Europe. It gives way, very often, to sterile discussions which still concern the so-called first generation GMOs (herbicide-tolerant plants, for example, developed in the 1980s), debates which do not take into account new developments in biotechnology research.

As plant pathologists, we would like to be able to have plants that are resistant to pests, in order to reduce the use of pesticides. Most pest-resistant plants, obtained throughout genetic manipulation available today (corn, cotton, etc.) are the result of researches carried out in extra-European countries and, as such, of course, they respond to plant health issues of important crops of those particular contexts. A larger investment of public money in Europe (and in Italy in particular) could have led to the development of pest-resistant cultivars of greater interest to our country (for example, tomatoes that are resistant to soil-borne pathogens, flavescence dorée resistant grapevine, apple varieties resistant to scab...). Unfortunately, we do not have any crops which, thanks to these characteristics, could be of interest to our agriculture. Today, moreover, new methodologies are used to obtain pest resistant plants: from a cisgenic approach (see box page 255) to genome editing (see box page 255).

Biotechnological applications aimed at producing stress-resistant plants, especially water-shortage resistant crops, are of great interest: having agricultural crops with lower water requirements might solve many problems due to drought in various geographical areas.

Interesting applications are also those that have led to the production of genetically modified plants with a higher nutritional content. Many of our common food crops are, in fact, not perfect from a nutritional point of view: their composition and

M. L. Gullino, *Spores*,
https://doi.org/10.1007/978-3-030-69995-6_56

their content in proteins, starch and oils, as well as in vitamins and micronutrients, might be improved to obtain healthier food. Interesting examples are potatoes with higher starch content, which absorb less fat during frying. Modified rapeseed and soya with a higher level of oleic acid (monounsaturated fat) produce oil similar to groundnut and olive ones and are marketed in Australia, Canada, USA. Turnip has been modified to produce oil similar to coconut and palm oil and is used in Canada and the United States to coat chocolates, in food glazes, and in cosmetics.

Very famous among the so-called second generation GMOs, is the "golden rice" enriched in vitamin A, whose name comes from its yellow colour. This rice has an increased production of beta carotene, a precursor of vitamin A, in its edible parts. Perhaps less known but just as useful as a fortified food is rice enriched with iron (see box page 256).

As already mentioned, public opinion is not fundamentally opposed to the so-called "red biotechnologies" used in biomedical field, as it considers them useful for human health. Unfortunately, few are aware that green biotechnology as well can be of great use to human health, as it works on plants in order to provide vaccines and more nutritional cultivars. These applications are of particular interest especially for developing countries (see box page 257).

In addition, genetically modified plants can produce proteins with pharmaceutical properties (enzymes, growth hormones) or biodegradable polymers (elastin, collagen) with possible industrial applications.

All biotechnological applications must be assessed for their effect on human health and environmental safety during production, and a careful assessment of these risks must be carried out in all cases. Tools and technologies are now available to carry out these studies and present research methods are able to evaluate the risks associated to them.

More public investment is essential for a more correct approach to biotechnology in order to enable the issues of interest to growers and consumers to be tackled in this sector, assessing risks and benefits from time to time. A more participatory approach, involving the most interested stakeholders, might help overcome old obstacles.

This would also allow the debate on a subject of vital importance for the future of worldwide agriculture to take place in a more relaxed way in the future, using the scientific data made available by research.

Going to depth

From *trans* to *cis*

Since the presence of exotic genes in our food is perceived with alarm for its real or presumed risks for environment and consumer health, and highly discussed for ethical reasons, a cisgenic approach has been developed to improve pathogen resistance and crop quality traits, for example in the case of apple tree, transferring genes from a compatible donor plant.

Going to depth

Genome editing: a new much softer technique

Genome editing (also called gene editing) is a group of technologies that give scientists the ability to change an organism's DNA. These technologies allow genetic material to be added, removed, or altered at particular locations in the genome. Several approaches to genome editing have been developed. The best known is CRISPR-Cas9 that has generated a lot of excitement in the scientific community because it is faster, cheaper, more accurate, and more efficient than other existing genome editing methods. CRISPR-Cas9 was adapted from a naturally occurring genome editing system in bacteria. The bacteria capture snippets of DNA from invading viruses and use them to create DNA segments known as CRISPR arrays. The CRISPR arrays allow the bacteria to "remember" the viruses (or closely related ones). If the viruses attack again, the bacteria produce RNA segments from the CRISPR arrays to target the viruses' DNA. The bacteria then use Cas9 or a similar enzyme to cut the DNA apart, which disables the virus. The CRISPR-Cas9 system works similarly in the lab. Genome editing is of great interest in the prevention and treatment of human diseases.

Different types of risotto

Vitamin A deficiency causes partial or total blindness to 500,000 children/year: most national authorities resort to expensive food integration programmes. Traditional genetic improvement techniques proved to be ineffective in producing rice with high levels of vitamin A. Golden rice has been obtained by researchers of the Swiss Federal Institute of Technology of Zurich by introducing three genes into rice, two from narcissus, one from a bacterium. Thanks to an agreement with a private company, Golden rice technology has been used by several research centres in India, China, Philippines, where crossbreeding programmes have been carried out with local varieties, able to adapt to local habitats.

Iron-enriched rice, thanks to the addition of the bean ferritin gene, contains twice as much iron in the endosperm. Iron deficiency causes anemia in pregnant women and children in Asia and Africa. The use of iron-enriched rice might solve this problem.

Tasty vaccines

Edible vaccines production seems to be very promising, especially for its use in developing countries, where traditional vaccines are often too expensive and have not few practical contraindications: they must be stored at low temperatures and administered by specialists, using costly material (needles, for example). Producing plant-based vaccines and drugs reduce costs and make them cheap enough for the most deprived people. In addition, plants producing vaccines can be an alternative source of income for farmers in developing countries. Edible vaccines are developed by introducing antigens of human pathogens into plants. Eating fruits of such plants, like a real vaccination, induces the production of antibodies that trigger an immune response to certain diseases. Research in this field has allowed for example to obtain modified bananas for the induction of antibodies against hepatitis B. Bananas and potatoes have also been developed as a protection against gastrointestinal diseases. Antibodies against cancer, expressed in rice and wheat caryopsis, allow the detection of cells of lung, breast and colon cancers. Applications in this field are useful for both diagnostic and therapeutic purposes and might have a huge potential to solve infectious diseases in developing countries.

Dialogues

There are some people, absolutely sane (or presumed such), that have always been convinced of the importance and usefulness of communicating with plants. Discovering how plants themselves interact with each other and with their pests is even more fascinating. This dialogue, in fact, does not imply words, but an exchange of signals through chemical compounds. In particular, it would be some volatile components (VOCs) to intervene in the communication processes. For example, when *Artemisia* species are damaged by insects, they release into the air a bouquet of VOCs, consisting of highly volatile substances [ethylene, methanol, methacrolein and some monoterpenes (A)] and heavier compounds, with lower volatility [cis 3-hexanal (B) and trans-2-hexanal (D), methyljsmonate (C) and oxygenated monoterpenes (E)] and through these fragrant messages warn wild tobacco plants growing around, triggering defense strategies that protect them from insects.

Microorganisms also produce a wide range of volatile substances (VOCs) belonging to different chemical classes (aldehydes, alcohols, esters, lactones, terpenes and sulphur compounds) characterized by low molecular weight and high volatility. The inhibitory effects of VOCs produced by microorganisms on pathogenic bacteria and fungi are of particular interest for biological control strategies. For example, emission of VOCs, especially sesquiterpenes, has been found in the case of *Fusarium* antagonists.

Recent studies have shown that plants are able to emit sound vibrations and that their "words" produce visible effects on the growth and health of others (see box page 260). In short, basil somehow "speaks", while peppers "listen". It seems that plant can even "eavesdrop" on the conversations of their neighbours, and that what they "listen" may influence their growth—in either a good or a bad way, depending on the neighbours (see box page 261). According to a study carried out at the University of Western Australia, plants would use acoustic signals in order to communicate with each other. Their finding would suggest that plants not only can "smell" chemical substances and "see" the light emitted by their neighbours (communication modes already known to scientists), but that they can even "listen to the sounds" emitted by other plants. Incredibly, the same result was achieved after wrapping the plants in black plastic sheets, so that they could not exchange radiations or chemical signals.

© The Author(s), under exclusive license to Springer Nature Switzerland AG 2021   259
M. L. Gullino, *Spores*,
https://doi.org/10.1007/978-3-030-69995-6_57

One way or another, the chilly seedlings were able to detect what kind of plants their neighbours were, and react accordingly. The simplest, and perhaps the most intuitive hypothesis, is that what is exchanged is a sound vibration, as sounds propagate very well through various means. Some cases of interactions among plants and humans sound as miracles (see box page 262). Nowadays, the new technologies can indeed help (see box page 263).

Plants have feelings too

In 1966 Cleve Backster, a specialist with the CIA, best known for his experiments with plants using a polygraph instrument, accidentally discovered that the *Dracaena* he had in his office reacted almost "emotionally" to what was happening around. It all began one day, when he wanted to measure the electric conductivity variation of its leaves as a result of watering. Backster watered the potting soil and shortly afterwards measured the electrical conductivity on *Dracaena* leaves. He expected an increase of it; instead, the value of this parameter decreased as it usually happens in a human being who feels a slight emotion. Very strong responses occurred when live shrimps were dropped into boiling water in the presence of the plant. Backster then performed other experiments and realized that the plant reacted inexplicably to certain stimuli, such as the presence of people who, in front of it, had previously injured other plants or small animals, showing electrical responses indicative of excitation, something he interpreted as an expression of "fear". It seemed in fact that the plant remembered the identity of those who had injured other plants or small animals in its presence: what if this ability to store information could be considered an elementary form of "memory"?

What Backster's experiences showed is that plants are able to manifest physical and chemical responses (now called "Backster's effects") similar to human biological states commonly defined as "emotions".

Plant perception

In 1966, while he was researching electrical responses in plants being watered, Backster connected a polygraph to one of the plant's leaves he was working on and, to his great surprise, found out that the polygraph recorded fluctuations in the electrical resistance quite similar to those of a man undergoing a truth test.

Was it possible that the plant was experiencing some kind of stress? If, for example, he had burned one of its leaves, what would have happened? Just as he was thinking these things, the polygraph needle went crazy, suddenly reaching its peak. Backster was convinced that the plant somehow had felt his intentions: it was sentient!

The miracles of Don José

Getting 100 lb cabbages, 5 feet long beet leaves, corn plants that reach a height of 16 feet, 150 tons of onions per hectare instead of the usual 16 tons and so on... The person who manages to perform these prodigies with environmentally friendly techniques is Don José Carmen Garcia Martinez, in the Santiago Valley at the centre of Mexico.

Legend has it that in the 1970s, in order to cope with top soil depletion, the young farmer did not despair, and instead of using fertilizers and chemical products, he began to talk to the soil. According to him, earth and plants have a form of intelligence that allows them to communicate with humans. Just know how to talk to them and above all to listen to them! This farmer would be able to grow anything without using pesticides and with homeopathic doses of fertilizers, obtaining, in saline soils, disease-resistant plants. When asked how he manages to achieve such wonders, Don José Carmen explains his method: "Plants have a life like every human being and animal. We have to learn to know them, to treat them gently. Plants do understand, they know. Not all people have affinities with plants, and not all plants have affinities with humans. It is a matter of compatibility, like people's blood types. Plants themselves can be grouped according to affinity, depending on their energy".

An app for talking to plants

Some people say that if you want to have healthy plants, you need to talk to them. But what if the plants themselves were able to talk to us and say what they need? On the occasion of CES (Consumers Electronic Show) 2013, the most important event dedicated to the world of technology and home automation held every year in Las Vegas, was presented a sensor that allows plants to communicate through an application specifically designed for smartphones. It works very simply: it sends pulses via bluetooth to indicate the amount of water or light our plants need, and inform us on the level of humidity and fertility of soil required by each species, whose peculiarities are collected in a database developed by a team of botany experts. The application is updated every 15 min to give us a precise picture of the state of our favourite plants.

Why Plant Diseases Still Scare the World?

This book is not a treatise on plant pathology and so I do not want to dwell on too technical aspects. However, with a few examples, I would like the reader to understand how some diseases that have caused famine in the past are still potentially dangerous. Today the increased knowledge in biology and epidemiology and the availability of effective control means simply allow to manage plant diseases avoiding huge yield losses. Nevertheless, we need to keep the focus on any phytopathological issues. A plant pathologist must always be alert, as I will clarify through some examples, mainly related to cereal diseases, which are still likely to cause very serious losses, also due to the importance that cereals have in our diet. At present, at world level, contrary to what many of us think, barns are not at all full of grains and other agricultural products. The "Arab spring", which caused riots in many Arab countries of the Mediterranean basin in the years 2010–2011 and consequent political upheavals, teaches what may mean sudden increases in the price of staple food.

An example is given by the three most important rust diseases of wheat, commonly indicated by reference to the colour of leaf pustules (organs filled with their spores): black rust, brown rust and yellow rust. Today, at least in the industrialized countries, wheat rusts no longer cause the devastations of the past, thanks, above all, to the use of resistant varieties. The search for such varieties, as we have seen before, is often a continuous run-up between geneticists and the pathogen often able (this is the case of causal agents of wheat rusts) to change their genetic patrimony, differentiating new races capable of attacking varieties previously considered resistant. A recent example of how it is impossible to sleep soundly is represented by the sudden appearance, in 1999, in Uganda, of a new strain of *Puccinia graminis* (causal agent of black rust), called Ug99, extremely virulent and able to attack all presently cultivated wheat varieties. This new strain of black rust has spread very quickly in various countries of central-eastern Africa and, thanks to its spores carried with great speed by the wind, it has reached the Near East and central-southern Asia, causing very serious losses. All European and North American cereal-producing countries immediately became alarmed, fearing the

M. L. Gullino, *Spores*,
https://doi.org/10.1007/978-3-030-69995-6_58

introduction of this new race of *Puccinia graminis*. Today, moreover, trade glob-alization and the speed of transportation simplify to the maximum the rapid spread of new pathogens or, as in the case of Ug99, of new races.

Therefore, after millennia, wheat rust can still cause considerable concern, forcing the adoption of severe preventive measures.

Still talking about cereals, another remarkable disease is common bunt of wheat, caused by two *Tilletia* species. Pathogen fructifications are formed in place of the caryopsids (seeds), of which they keep the walls. At threshing, the walls break and black clouds are formed that rise to the sky, emanating an unpleasant stink of rotten fish. At the end of the last century, in the presence of environmental conditions very favourable to the development of the pathogen, it caused serious attacks of wheat bunt in the United States of America, with serious yield losses and huge economic consequences, because exports were stopped and it was necessary to carry on very expensive eradications. Today *Tilletia* is still considered a very dangerous pathogen.

Another example of how a pathogen can cause devastating damage is the recent bacterial canker of kiwifruits. In this case, a bacterium already present for some years starts to differentiate more virulent strains and spreads through plant material in different geographical areas. This disease is particularly difficult to control, because the use of antibiotics is forbidden. Kiwifruit growers in many countries had to explant the trees as an extreme remedy. On the other hand, in many production areas, they face serious problems in choosing the varieties to be planted. The problem will be probably solved with the development of resistant cultivars, not an easy task, however.

Very similar, in some ways, is the case of *Xylella fastidiosa*, the bacterial disease that is interesting olive trees in the Salento area of Southern Italy: spread by an insect, it is a new pathogen for European olive plantations, although already present in other geographical areas and on other species. Being antibiotics banned, also in this case chemical control is difficult, and mostly limited to the control of its vector. Also for this reason, *Xylella*'s olive quick decline syndrome is devastating this very traditional and typical crop of the Mediterranean regions. It has been estimated that, if *Xylella* reached other European olive production areas, the damage could reach an average of 50% yield losses, causing the loss of 300,000 workplaces.

Agriculture of the Future: What Role for Plant Protection?

Paradoxically, protecting crops from pests is more difficult today than it was in the past, despite the availability of effective and sophisticated technologies, and it will be even more so in the future. In a context of general crisis that requires to lower production costs, and with the need to limit the environmental impact of interventions, the role of research will be increasingly important.

Crop protection will have to take more and more account of the globalized trade effects in the future: plant material, which is increasingly produced in a few specialized companies, often in developing countries, is often vectoring several pests that, within a few weeks, may spread in very distant geographical areas (see box page 268). Many of the most serious and important pathogens that have infected Italian horticultural crops in recent years moved through infected seeds or propagating material (see box page 269). To prevent this phenomenon, it is essential to have highly trained technicians and rapid, accurate diagnostic methods, shared among labs. New and old pathogens—and at times re-emerging ones—affect and will affect our crops even more frequently than they have done so far, and the emergence of new plant diseases will be amplified by the ongoing climate change.

In the coming years it will be very interesting and important to develop strategies to mitigate the effects of climate change, also in relation to the stress caused to plants by the new environmental conditions (see box page 269) and to the spread and severity of certain diseases, also monitoring how it affects the management strategies themselves.

The use of chemical control means will be increasingly complicated by the limited availability of registered pesticides as a result of the implementation of more and more stringent regulations. It is to be expected that there will be a number of critical situations in the future (soil disinfestation, minor crop disease management, post-harvest protection, etc.), with significant economic consequences for growers. With so many critical issues, it is essential that research focuses on the most relevant topics, in a global perspective: crop biosecurity, surveillance on plant material, development and sharing of innovative diagnostic methods, commercial development of products alternative to pesticides, are just some of them. An

M. L. Gullino, *Spores*,
https://doi.org/10.1007/978-3-030-69995-6_59

effective training system, able to adapt to current, increasingly complex and evolving scenarios, will have to come along with research.

The challenge for researchers who work at the service of farmers will be meeting their expectations, providing extension services with the most appropriate solutions (see box page 270).

Watch out for aliens!

Although it is impossible to predict how many new pests will come to plague our crops in the future, it is easy to figure out that in an increasingly globalized world more and more pathogens will be moving at great speed from one continent to another at a speed proportional to that of transport means.

These newly introduced pests are now called "alien species" or "invasive species"—aliens, in short—as they come from far away countries where, frequently, they have found forms of coexistence with their hosts. When they get into a new geographical area, they sneakily attack their host plants, causing sudden damage. Until they are properly detected and identified, and until control means are found, it is real trouble for farmers.

Beauty treatments for seeds

All in all, the various pathogens do nothing but induce a stress situation in the affected plants, very similar to what sometimes happen to us. How will sick plants be treated in the future? The tendency to use control means alternative to chemicals might lead to a different future for plant disease management. Who knows that, in less serious cases, might not be possible to identify soft treatments similar to those offered in a spa? After all, kind of "massages" with essential oils and plant extracts are already used for seed dressing and post-harvest treatments, for example. In the future, it might be possible talking about real "seed massages".

Burned out plants?

What happens when the stress is more severe? We have already talked about the plant neurobiologist. Will we take plants to the psychologist? Will we treat them with hypnosis? When I ask my friends who are psychologists how they manage to detect the slightest signs of malaise in their patients in order to treat them, they answer: "Exactly as you do with plants". Perhaps plant pathologists are already kind of psychologists who use their sensitivity to "understand" plants and catch their disease symptoms. Why not? It might be fascinating trying to imagine new professionalism in this sense.

Space plant pathologists, with feet on the ground

What will future plant pathologists look like? For sure, they will have more sophisticated tools and will be able to better understand how and why plants get sick. They will work in different environments—who knows, maybe even in space! In any case, they will have to keep their feet on the ground. Perhaps in this answer lies the essence of their work. Thanks to their sensitivity, they manage to catch plants' malaise and treat them with different means, but always with passion. In other words, passion is what will always guide plant pathologists.

Emerging Disciplines

Communication between different organisms is always complicated by the difficulty of synchronizing the intentions linked to the signal emission with the response it generates. This is even more complicated when communication channels are established between very distant organisms, such as plants, animals and fungi, because in this case there is often a considerable difference between the organism that emits the signal and the one that receives it (and the interpretation of it). Plants emit signals to overcome their lack of mobility that creates a physical obstacle to their potential of acquiring resources, defending themselves from predators, dispersing their genes. This obstacle is usually overcome with the development of mutual relations with other organisms through specific communication channels.

While in the past, much interest was placed on the evolution of animal communication, in more recent times the exploration of signalling between plants and at all levels of biological organization has attracted increasing attention, thanks also to the finding of new and unexpected potential of plants, now seen as information processing organisms with complex forms of communication.

According to some, plant neurobiology represents a new promising crossover research discipline. Among its passionate founders, there is Stefano Mancuso, director of the International Laboratory of Plant Neurobiology (LINV) of the University of Florence, where the effect of sound vibrations on plants is studied (see box page 273).

Communication is definitely one of the key words of plant neurobiology. The cells of a single plant communicate with each other in similar ways to those so far considered exclusive to animals.

Plants, however, are also very skilled in communicating with other organisms and species. Their roots, for instance, secrete in the soil a great quantity of substances that act as true signals, and the same do leaves and flowers with volatile molecules. In some cases, they are "chemical weapons", directed against the surrounding plants with the aim of hindering their growth and development, or against predators, to warn them off. Other signals, instead, are "friendly", such as those used to attract pollinators or alert other plants of their community of the presence of

M. L. Gullino, *Spores*,
https://doi.org/10.1007/978-3-030-69995-6_60

dangers. Several studies, for example, have demonstrated that plants attacked by herbivorous insects or pathogens emit volatile substances that can signal the danger to nearby plants, giving them time to prompt their defence strategies, usually through changes in their physiology that make them more resistant. Many plants attacked by predators or pathogens produce substances that are repulsive against the enemy, or able to attract predators of the enemy itself (according to the ancient doctrine stating that "the enemy of my enemy is my friend"). Among the most common plants that adopt this strategy, there are tobacco, tomato and eggplant. These abilities might be exploited for agricultural purpose: if we flood a crop with a "warning message", we are preparing it to face an attack, inducing it to activate its resistance strategies.

Forgetting plants or considering them secondary organisms is too easy, since they do not move and we do not hear them. However, this is wrong. In fact, plants do move! Simply, their potential is masked because they operate on much lower time scales than animals, and unlike the latter, that move into space, they stay "on the spot" (just think of seedlings that orient their position and growth in order to maximize their contact with sunlight), and usually move extremely slowly—with some exceptions, such as carnivorous plants, or *Mimosa pudica*, which closes the leaves very quickly at the least contact. As for the fact that "they do not make themselves heard", much depends on our ability of "listening". Plants, in fact, have a very complex communication system, based on a great variety of molecules (amino acids, sugars, secondary metabolites, volatile substances) with which they "converse" with their neighbours or with animals.

In the last years, an internal system of information transmission has been discovered at root level, a plant nervous system which can be considered analogous to that in animals (see box page 274). In fact, as demonstrated by neurobiology researchers at the University of Florence, Italy, in plants there is not a "physical" structure comparable to nervous tissue, consisting of neurons and other nerve cells specialized in the transmission of electrical signals. In other words, plant do not have neurons, but some of their cells—those of the root-apical zone, that is at the tip of the root—are able to deliver electrical signals in the form of action potentials (variations in the potential difference between the inside and the outside of the plasma membrane) and transmit them to nearby cells (see box page 275).

We should remember that already Charles Darwin, in his book *The Power of Movement in Plants* (1880), stated that root apex acted like a "diffuse brain" in plants, able to "receive impressions from the sense-organs" and to direct their movements as they were deciding on the strategies to follow. Today we know that roots also have mechanisms for the processing and transmission of these signals. Many neurotransmitters present in our brain (glutamate, serotonin, dopamine, acetylcoline, etc.) have been found in plants as well. In this case, they are not called neurotransmitters, because they are not in a brain and because their function is not always known, but they are a kind of plant command centres, and for some of them a fundamental role has been shown in the mechanisms of information transmission.

For instance, a root has the constant need to know with extreme precision what happens in the surrounding environment. This "knowledge" derives from the activity of the root apices, each of which is able to "feel", that is perceive and evaluate, at least 15 different chemical and physical parameters, such as temperature, salinity level, humidity and so on, which have to be integrated and developed to identify the optimal growth direction. It has been discovered that glutamate plays an essential role on this processing: if it is lacking or is present in excess, the root acts as if it had lost the sense of orientation, and grows abnormally.

This research is carried out using *Arabidopsis thaliana* (a small and versatile plant of which many details are known), tobacco, corn, tomato, olive and grapevine.

Musical rows of vines

Walking among the vineyards of some wineries in Montalcino, Tuscany, home of the famous Brunello, one of the Italian finest wines, it may happen to find ourselves immersed not only among grapevines, but also in music: Mozart, Vivaldi, baroque works. It is not an extravagance, but a scientific experiment and among the researchers who work on it there are also those of Stefano Mancuso's group. The meaning of these experiments carried out in Montalcino, is understanding what happens to a plant subjected to sound vibrations of particular frequencies: by the way, the choice of the musical repertoire has no scientific value, it is simply a matter of "taste". The first results are coming: "It seems that wine grapes exposed to music ripen about ten days before the others: an interesting behaviour for producers, who fear the bad weather in case of late harvest". Other research suggest that sound vibrations could protect plants from insects, interfering with their reproductive behaviour.

Root apices as plant command centres

The turning point that led to understand the central role played by root apices in plants was the discovery of a particular zone of plant roots—called "transition zone"—which consumes much more oxygen than nearby areas, a condition that is a sign of strong energy demand and, therefore, of the presence of some intense activity. Yet, at the beginning, the transition zone did not seem to be involved in high energy demanding tasks, such as cell division. So it was logic to wonder why the transition zone consumed so much oxygen if, apparently, it was doing nothing special. The hypothesis was that it was involved in an activity analogous to that of neurons and in fact, as researchers of the University of Florence confirmed, the cells of this area focus all their resources upon perception of environmental signals and transmission of action potentials.

At the roots of plant intelligence

One of the most obvious reasons explaining why plants have developed a neural-like activity at the level of root apices, is their underground origin: soil, in fact, is a more stable environment than the atmosphere for what concerns temperature and humidity, as well as being protected from animal predation and ultraviolet radiation. As for "cognitive" tasks, we have already mentioned some of them, such as the ability to collect and integrate environmental information and react accordingly. Moreover, plants show also effective processing abilities, such as intra-species and inter-species communication, and a learning behaviour connected with memory mechanisms and cost-benefits calculation.

2020: The International Year of Plant Health

As we have learned throughout this book, plants are under constant attack from invasive pests. These pests can severely damage crops, forests, and other natural resources that people depend on. Every year, they cause billions of dollars of losses in crops and trade revenue, in addition to expensive eradication efforts. People, especially through international travel and trade, most often spread them. Despite declining resources for research in general, and for agricultural research in particular, Universities and research centres as well as international, regional and national plant health organizations continue in their efforts to protect plant.

Plants are the foundation of life on earth. They produce the oxygen we breathe. They provide more than 80% of the food we eat. We use them to make clothes, shelter, medicines, and many other things that are essential to our lives. For nearly half of the earth's population, plants are a primary source of income. Almost every country trades plants and plant products to create wealth and support economic development. A threat to plant health is also a threat to the health and prosperity of people across the globe—especially the most vulnerable.

The UN Food and Agriculture Organization and the International Plant Protection Convention Secretariat, based at FAO, welcomed the UN General Assembly's adoption today of a resolution proclaiming 2020 as the International Year of Plant Health (IYPH) (see box page 278). This celebration increased awareness among the public and policy makers of the importance of healthy plants and the necessity to protect them in order to achieve the Sustainable Development Goals. Any effort to achieve the vision set out by the 2030 Agenda for Sustainable Development must acknowledge the critical importance of plant health. The International Year of Plant Health helped raising awareness, driving concrete action and ultimately contributing to a safer, more prosperous and peaceful world.

All over the world Institutions and plant pathologists did plan many, diverse outreach activities. The Covid-19 pandemics did partially affect such activities, but, in the meantime, it emphasized once more, if still needed, the importance of plant health.

© The Author(s), under exclusive license to Springer Nature Switzerland AG 2021
M. L. Gullino, *Spores*,
https://doi.org/10.1007/978-3-030-69995-6_61

Ralf, the father of the international Year of Plant Health

The idea of a full year devoted to plant health celebration is due to Mr Ralf Lopian, Senior Advisor at the Ministry of Agriculture and Forestry of Finland. He designated this year in order to increase awareness about the importance of plant health, to promote and strengthen national and international projects aiming to reduce plant diseases caused by climate change, and to encourage dialogue between different relevant actors. He is sure that after such a year, stake-holders, decision-makers, consumers and everyone else will know more about plant health and the related threats, and will better understand the importance of plant health.

Plant Pathologists and Plant Pathology at the Time of the Corona Virus Pandemic

Italy has been, after China, the first country being severely affected by the coronavirus before other European countries and the United States started to be interested by the pandemic. At the beginning, the country has been for a while considered the black swan. Everybody know how much Italian people like the good life, are socially active, embrace and kiss each other. Being forced to stay apart was difficult and shocking, at the beginning. What we saw happening in China in the Hubei province looked so distant. Sudden, we found ourselves in the same situation. With Europe treating us as infected by plague. Some countries started closing their borders. Italians were no more accepted in many other European countries. However, human and animal pathogens, as well as plant pathogens, do not have borders. Plant pathologists know this very well, and are able to understand how an epidemic can act. In Italy, we just suffered the *Xylella* epidemics on olive groves. A somehow similar story, with plants instead of humans. In a few weeks, other European countries started to be severely affected and soon after the coronavirus epidemics spread all over the world and in some countries. At the time of writing, the second wave starts being under control in some areas and a third one is expected. There is some understanding that this pandemic will start to be under control only after the vaccination campaign will be on place. As plant pathologists, also our work life has been affected. Universities have been (and still partially are) locked down, many labs were closed, activities have been kept at a minimum, just to maintain plants and cell cultures alive. Many International and National Congresses have been cancelled or postponed. Travels stopped, airline companies are no more connecting most countries. Our daily life changed. Everywhere we had to adapt very quickly (and we did) to distant teaching. Most courses have been provided this way, with all the problems related to the practical part of our discipline. Most researchers and technicians had to shift to smart working, not very popular before this event. It was sad walking throughout empty labs and greenhouses, though. The Covid-19 did slowdown of course our research; many grant calls were postponed. Field trials for the 2020 season were affected. As plant pathologists, we are particularly interested in better understanding the reasons why

M. L. Gullino, *Spores*,
https://doi.org/10.1007/978-3-030-69995-6_62

Italy has been so severely affected. Because of the presence of an aged population? Italy is second after Japan, in terms of longevity. Has environmental pollution, particularly high in the Po Valley, had a role? Why the methodology applied to test people has not been validated and applied in the same way all over Europe? Why the standardized methods that we are using in our discipline were not applied in the different countries? Do we have enough strains of Covid-19 deposited in order to better understand the Italian population of Covid-19? Many of these questions are still open. The intensive research carried out in so many laboratories will soon provide answers. This epidemic, so devastating, will have very severe social and economic effects. Most of the scheduled celebrations for the International Year of Plant Health have been cancelled or deferred. Plant pathology, as a discipline, lost many opportunities to reach the public during this special year devoted, by the United Nations, to the celebration of Plant Health. However, being our discipline strictly connected with environmental health, should indeed take at least some advantage of such a negative situation to transform it into a *momentum*. Because plant health is strictly linked to environment, animal, human health. In a concept of circular health. As plant pathologists, we should indeed be able to let people understand the importance of plant health and its strict connection with environment, animal and human health.

Plant Health in the Frame of Circular Health: A New Vision

The Circular Health philosophy underlines the idea that the health of coexisting humans, animals, and plants in one environment is interconnected. Governing this interconnection is key to achieving the health of the whole system. Circular Health is a conceptual framework aiming at understanding the links and dependencies between the health of humans, animals, plants and the environment (see box page 282). It even overpasses the One Health concept, striving to bring all the different disciplines and perspectives towards the optimization of health outcomes for humans, animals, plants and the environment. A classic and simple example of the One Health approach involves diseases that are shared between animals and people, also called zoonotic diseases. In the past and still nowadays in most cases the one health concept only included human and animal health, leaving plants out. However, if still needed, the Covid-19 pandemic clearly showed the many interconnections among environment, plants, animals and humans. The interconnections are so many, and most of them are still unexploited. Antibiotic resistance, honeybees' decline, are just examples of problems that directly and indirectly affect at large health, food supply and environment. Artificial intelligence will be instrumental to the deployment of circular health, benefiting the future health for humans, animals, plants and the environment. As often happens in history, from a very negative experience, as the Covid-19 pandemic has been, a new and more interdisciplinary approach in research will develop, bringing promising insights. As plant pathologists, beside strengthening our connection with veterinary and medicine, we should indeed learn to change our focus from diseases to health. With an immediate need of a much more interdisciplinary approach.

M. L. Gullino, *Spores*,
https://doi.org/10.1007/978-3-030-69995-6_63

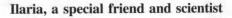

Ilaria, a special friend and scientist

Ilaria Capua, one of the brightest virologists in the world, is an Italian scientist leading the One Health Center at the University of Florida, US. The Centre aims at expanding areas of knowledge on health interconnections with the ultimate goal of improving health as a system. Ilaria is very well known for her past work on the Avian influenza and her very early efforts in promoting open access to genetic information on emerging viruses as part of pre-pandemic preparedness efforts. If the Circular Health concept will flourish, it will be thanks to her determination and vision.

Concluding Remarks

Hopefully this book, though very lightly, brought you in the wonderful world of plants and their health. Plants not only decorate the world, making it much more pleasant, but also provide us the oxygen that we breathe and the food and feed that nourish us as well as animals. That is why it is mandatory to love, respect them, also taking care of their health. Moreover, the Agencies in charge of supporting research should pay much more attention to the plant sector, understanding their contribution to health in general. Investing in plant health means investing in the health of the environment and in our future.

When your come across a sick plant – no matter if a tree or a flower – do not turn your head, looking for a healthy or nicer one. Think about the disease behind those symptoms: there is certainly someone who is worried about it and who is passionately taking care of that plant.

If plant health awareness is raised through this book and the desire to become plant doctor is awaken in some of you, it will mean I have reached my goal.

As far as I am concerned, retracing my life through the main theme of plant diseases has been very useful. Writing this book was not, despite my Calvinist attitude, wasting time that should have been devoted to research. It meant to me going back to my origin, to my enthusiastic period as a young fellow starting her research experience. I could fully realize how lucky I have been having had the opportunity to work in a wonderful environment, spending long periods abroad, with the best colleagues you can have, on so many challenging topics. Going throughout my nomadic life-style, I did also understand that it is probably the right time to store my too many suitcases, landing, finally, in the beautiful labs and greenhouses at Grugliasco. Time to share with my own territory the experience gained. Time to pass the witness to new generations of researchers who live in a much less fortunate period. What a wonderful trip!

Glossary

Active ingredient Component of a pesticide formulation that exerts a toxic action against specific plant pests.

Bacterium Prokariotic, unicellular organism, which multiplies by binary cleavage.

Biological weapons Agents capable of multiplying in the attacked organism causing death or disease in humans, animals or plants.

Canker Localized lesion usually in woody organs, resulting from the alteration of cortical tissues, which may be followed by hyperplastic processes of adjacent tissues.

Desiccation Rapid withering of large parts of leaves branches and buds.

Disease Any alteration affecting a living organism.

Epidemic Disease that takes on in a specific area, the character of a mass infection.

Epidemiology Study of the distribution and determinants of a disease in specified populations.

Epiphyte An organism that grows on the surface of a plant.

Eradication Intervention aimed at completely eliminating a plant pest.

Fumigation Method of pest control with pesticides acting as gasses.

Fungicide Chemical product able of inhibiting the development of fungal pathogens.

Fungi Eukaryotic heterotrophic, chlorophyll-free organism, consisting of hyphae.

Green Golf course area around the hole prepared to allow the ball to roll.

Hybrid Offspring resulting from the crossing of different species and varieties.

© The Editor(s) (if applicable) and The Author(s), under exclusive license to
Springer Nature Switzerland AG 2021
M. L. Gullino, *Spores*,
https://doi.org/10.1007/978-3-030-69995-6

Infection Process due to a plant pathogen invading host tissues with or without the appearance of symptoms.

Leaf layers Upper or lower surface of the leaf blade.

Lesion Area affected by a well-defined local alteration.

Mycotoxin Toxic metabolite produced by some fungi.

Mulching The exercise of protective covering of the soil around the plants with different materials.

Mycelium Interlacing of hyphae capable of performing the fundamental functions for the life of a fungus.

Outbreak The starting point of an infection or epidemic.

Pandemic An epidemic occurring worldwide or over a very wide area, crossing international boundaries and usually affecting a large number of people.

Parasite Animal or plant organism living at the expense of other organisms.

Pathogen Infectious agent capable of penetrating spreading and growing at the expense of host tissues

PCR (Polymerase chain reaction): a technique that allows the amplification of specific DNA traits by means of repeated replication cycles.

Pesticide Product used for pest control. Generally chemical, it may also be a microbiological substance (biopesticide)

Phloem Living tissue in vascular plants that transports organic nutrients from roots to leaves.

Phytoplasm Cell-wall free bacterium.

Phytotoxic Element or effect detrimental to the normal development and physiological functions of a plant.

Plant pathology Discipline studying plant diseases.

Plant protection The study of plant disease management.

Polyphagy Ability of an individual to live at the expense of different guests.

Quarantine Set of measures defined by law and decrees, aimed at excluding from entire territories organisms and/or plant material that might carry dangerous alien invasive species.

Resistance Inherent capacity of an organism to oppose the action of a pathogen.

Rhizosphere Mass of soil immediately surrounding the roots characterized by intense microbial activity.

Root anastomosis Connection between roots.

Saprophytes Microorganisms (bacteria fungi) able to proliferate on dead organic matter or wastes, unlike pests that attack tissues of living organisms.

Sclerotium Compact cluster of hyphae mostly round in shape, endowed with a particular resistance and able to maintain a long life even in presence of adverse environmental conditions.

Soil and substrate disinfection Soil treatment with physical methods (soil steaming or solarisation) or chemical means (fumigants) aimed at reducing the presence of weeds pests and diseases in soil.

Soil-borne Organism that lives in soil.

Spores Basic reproductive units of fungi.

Sporulation Spore production process.

Stomas (or stomatas) Pores found in the epidermis of leaves especially on their lower surface, of stems and other organs, which allow gas exchange.

Susceptibility Ability of an organism to become infected with a pathogen showing clear symptoms of disease.

Symptomatology Set of symptoms characteristic of a disease.

Systemic In a pesticide substance able to enter into the plant circulation in a pathogen the ability to distribute itself in all tissues of the plant.

Taxonomy Classification of plant and animal organisms.

Tolerance The plant's ability to withstand a pathogen infection by showing only mild symptoms.

Tracheomycosis Fungal disease affecting vascular tissues whose lumen is occluded by fungi and/or tillers or gums produced as a reaction by the plant itself.

Transgenic (plant) A plant in which one or more genes from different species either intact or modified, have been inserted using genetic engineering techniques.

Tumor Roundish enlargement of plant tissues due to the formation of hypertrophic and hyperplastic cells.

Virology The study of viruses.

Virulence Quantitative expression of pathogenicity.

Virus Submicroscopic infectious agent strictly intracellular, consisting of a single type of nucleic acid and proteins.

Wilting Irreversible weakening of tissues due to lack of water caused by biotic and abiotic factors.

Yellowing A symptom of infection caused either by pathogens, in particular certain viruses and phytoplasmas, or by a mineral deficiency.

Essential Readings

Capua I (2020) *Circular health: empowering the one health revolution.* Bocconi University Press
Dash M (2001) *Tulipomania: the story of the world's most coveted flower & the extraordinary passions it aroused.* Crown Publishing, New York City
Diamond J (1997) *Guns, germs, and steel: the fates of human societies.* W. W. Norton & Company, New York City
Mancuso S, Viola A (2013) *Verde brillante.* Giunti Editore, Milano
Matossian MK (1989) *Poisons of the past: molds, epidemics, and history.* Yale University Press, New Havem USA
Quammen D (2012) *Spillover. Animal Infections and the Next Human Pandemic.* W.W. Norton & Company
Schumann GL (1991) *Plant diseases: their biology and social impact.* APS Press, St Paul, MN, USA
Schumann GL, D'Arcy CJ (2012) *Hungry planet.* APS Press, ST Paul, MN, USA
Woodman-Smith C (1962) *The great hunger: Ireland 1845–1849.* Harper and Row

Specific Readings (for those of you who, by reading *Spores*, have become crazy for plant pathology)

Agrios GN (2005) *Plant pathology.* Elsevier Academic Press, Burlington, MA, USA
Horst RK (2013) *Westcott's plant disease handbook.* Springer, Dordrecht, The Netherlands
Schumann GL, D'Arcy CJ (2010) *Essential plant pathology.* APS Press, St. Pau, MN, USA